ANIMAL SCIENCE, ISSUES AND RESEARCH

THE ZOOLOGICAL GUIDE TO CRUSTACEA

ANIMAL SCIENCE, ISSUES AND RESEARCH

Additional books and e-books in this series can be found on Nova's website under the Series tab.

ANIMAL SCIENCE, ISSUES AND RESEARCH

THE ZOOLOGICAL GUIDE TO CRUSTACEA

NOELLE LACHANCE
EDITOR

Copyright © 2019 by Nova Science Publishers, Inc.

All rights reserved. No part of this book may be reproduced, stored in a retrieval system or transmitted in any form or by any means: electronic, electrostatic, magnetic, tape, mechanical photocopying, recording or otherwise without the written permission of the Publisher.

We have partnered with Copyright Clearance Center to make it easy for you to obtain permissions to reuse content from this publication. Simply navigate to this publication's page on Nova's website and locate the "Get Permission" button below the title description. This button is linked directly to the title's permission page on copyright.com. Alternatively, you can visit copyright.com and search by title, ISBN, or ISSN.

For further questions about using the service on copyright.com, please contact:
Copyright Clearance Center
Phone: +1-(978) 750-8400   Fax: +1-(978) 750-4470   E-mail: info@copyright.com

NOTICE TO THE READER

The Publisher has taken reasonable care in the preparation of this book, but makes no expressed or implied warranty of any kind and assumes no responsibility for any errors or omissions. No liability is assumed for incidental or consequential damages in connection with or arising out of information contained in this book. The Publisher shall not be liable for any special, consequential, or exemplary damages resulting, in whole or in part, from the readers' use of, or reliance upon, this material. Any parts of this book based on government reports are so indicated and copyright is claimed for those parts to the extent applicable to compilations of such works.

Independent verification should be sought for any data, advice or recommendations contained in this book. In addition, no responsibility is assumed by the Publisher for any injury and/or damage to persons or property arising from any methods, products, instructions, ideas or otherwise contained in this publication.

This publication is designed to provide accurate and authoritative information with regard to the subject matter covered herein. It is sold with the clear understanding that the Publisher is not engaged in rendering legal or any other professional services. If legal or any other expert assistance is required, the services of a competent person should be sought. FROM A DECLARATION OF PARTICIPANTS JOINTLY ADOPTED BY A COMMITTEE OF THE AMERICAN BAR ASSOCIATION AND A COMMITTEE OF PUBLISHERS.

Additional color graphics may be available in the e-book version of this book.

Library of Congress Cataloging-in-Publication Data

ISBN: 978-1-53616-366-7

Published by Nova Science Publishers, Inc. † New York

CONTENTS

Preface vii

Chapter 1 Cave Crustacean Decapods from Mexico 1
Luis M. Mejía-Ortíz

Chapter 2 Artemia Franciscana (Crustacea: Anostraca) in a Hypersaline Habitat in Abu Dhabi (United Arab Emirates) as Intermediate Hosts for Avian Cestodes 67
Rolf K. Schuster, Anitha Saji and Shaika Salem Obaid Al Daheri

Chapter 3 Behavioural Responses of the Non-Marine Ostracod Heterocypris iIncongruens (Crustacea: Ostracoda) to Natural Chemical Cues: Laboratory Observations 89
Dragana Miličić, Tatjana Savić, Branka Petković, Milica Potrebić, Jelena Trajković and Sofija Pavković-Lučić

Chapter 4 A Zoological Overview of the Larvae of Different Groups of Lobsters 113
Brady K. Quinn

Index 165

Related Nova Publications 171

PREFACE

The Zoological Guide to Crustacea opens by providing an update on the cave crustacean decapods from Mexico, because in the last thirty years several species have been described. Mexico has diverse cave environments according to cave origin, such as: karstic caves, volcano caves with freshwater habitats and anchialine caves with marine and brackish habitats.

The following study focuses on a total of 2,700 Artemia franciscana collected at the Al Wathba Reserve between December 2017 and May 2018, which were examined for the presence of larval helminths by direct microscopy of glycerin mounted specimens. Of these, 341 (= 12.6%) specimens contained different species of the Hymenopepididae and Progynotaeniidae families.

The penultimate study investigates whether the non-marine ostracod Heterocypris incongruens can detect and react to chemical compounds derived from a predator and from injured conspecifics.

The closing study demonstrates the diversity of larval phases and their importance to lobster populations and fisheries, and provides a summary of larval developmental patterns and behaviors and their potential impacts on larval dispersal and lobster demographic connectivity.

Chapter 1 – This chapter is an update of the cave crustacean decapods from Mexico, because in the last thirty years several species have been described. Mexico has diverse cave environments according to the cave origins. There are karstic caves, volcano caves with freshwater habitats and anchialine caves with marine and brackish habitats. The decapods are a conspicuous taxonomic group, and in the caves they have colonized these environments. In Mexico, the following cave species have been exclusively registered; Procarididae (1), Anchialocarididae (1), Atyidae (4) Agostocarididae (2), Alpheidae (3) Barburiidae (1), Hippolytidae (3) Palaemonidae (10), Cambaridae (5), Pseudothelphusidae (5) and Trichodactylidae (4). This chapter include taxonomic keys and a compilation of brief descriptions of each species and their notes on the habitats and distribution on the karstic regions in Mexico.

Chapter 2 – Crustaceans of the genus *Artemia* inhabit hypersaline aquatic biotopes. They are economically important as their larval stages are used in aquaculture as food for fish and crayfish larvae. Serving as food source for waders and charadriform birds *Artemia* spp. play a role as intermediate host in the life cycle of avian cestodes. The Al Wathba Wetland Reserve is a complex of natural and man-made water bodies about 40 km of central Abu Dhabi (United Arab Emirates). Founded in 1998 it has been recognized by the Convention of Wetlands of International Importance in 2013. In 2018, the wetland was listed in the IUCN Green List of Protected and Conserved Areas (UNEP-WCMC 2018). Apart from a stable population of greater flamingos more than 260 other bird species can be spotted there. Between December 2017 and May 2018 a total of 2,700 *Artemia franciscana* collected at the Al Wathba Reserve were examined for the presence of larval helminths by direct microscopy of glycerin mounted specimens. Of these, 341 (= 12.6%) specimens contained cestode cysticercoids of eight) different species of the Hymenopepididae (*Flamingolepis liguloides, F. flamingo, Wardium fusa, Confluaria podicipina*), Dilepididae (*Eurycestus avoceti, Eurycestus* sp., *Anomotaenia tringae*) and Progynotaeniidae families (*Gynandrotaenia stammeri*). There was no significance in cysticercoid prevalence between male and female hosts. Flamingo specific species, *F. liguloides* and *F. flamingo,* were the most frequently found species. The number of cysticercoids varied between one and twelve and up to three species were detected concurrently in one shrimp. In addition, one shrimp contained a spirurid larval stage.

Chapter 3 – In aquatic environments, chemical cues are very important for the perception of danger, especially when visibility is low. It is known that ostracods rely on chemical senses to detect predators, which is essential for survival. Occurrence of alarm signals in the surroundings can affect their activities and behaviour. The present study investigates if the non-marine ostracod *Heterocypris incongruens* (Crustacea: Ostracoda) can detect and react to chemical compounds derived from a predator (*Triturus* spp. larvae) and from injured conspecifics. Also, the study aims to investigate whether habitat-substrate selection is at play when the individuals are in potential danger. The obtained results indicate that predator-derived chemicals and conspecific alarm cues induce specific behavioural responses: forming of aggregations, reduction of locomotion or camouflaging. It is possible that *H. incongruens* rapidly evaluates environmental cues and modifies defensive strategies depending on the type of semiochemicals perceived.

Chapter 4 – Marine decapod crustaceans that crawl ('Reptantia'), brood their eggs (Pleocyemata), and are not considered 'crabs' or 'shrimps' are generally termed 'lobsters'. These include members of 6-7 different decapod infraorders, including the clawed nephropoid and reef lobsters (Astacidea: Nephropoidea and Enoplometopoidea, respectively), blind lobsters (Polychelida), ghost or mud lobsters ('Thalassinidea' = Axiidea and Gebiidea), clawless spiny, slipper, and furry lobsters (Achelata: Palinuridae, Scyllaridae, and 'Synaxidae,' respectively), and the 'living fossil' glypheid lobsters (Glypheidea), as well as certain crabs that are also commonly termed 'squat lobsters' (certain members of Anomura: Chirostyloidea and Galatheoidea). Many lobsters support important fisheries, and they play important ecological roles as large, benthic consumers in marine communities. All lobsters develop through a pelagic (planktonic and/or pelagic) larval phase, which differs considerably from the benthic juvenile and adult phases of their life cycles in terms of its morphology, behavior, physiology, and ecology. The larval phase is vital to the maintenance of lobster populations and fished lobster stocks as it supplies new recruits to benthic populations, but for some taxa relatively little is known about their larval development. This chapter briefly reviews the larval phases of all major lobster taxa, with particular emphasis on the variations in larval developmental patterns, morphology, life cycle characteristics (duration, type and number of phases and stages, etc.), and behaviors while in the water column across taxa. Lobsters develop through two or three larval phases after hatching that are comprised of one to several different stages each. A short-lived 'pre-zoea' stage has been observed after hatching in nearly all lobster taxa, but its role and status within the life cycle remains unclear. This is followed by varying numbers of zoea (mysis) stages after hatching, which in some cases are highly modified, as in the phyllosoma larvae of the Achelata and the 'eryoneicus' larvae of the Polychelida. The zoeal stages are then followed by a decapodid or 'postlarva' stage, which is a strong swimmer in some taxa and eventually settles to the benthos. This chapter demonstrates the diversity of larval phases and their importance to lobster populations and fisheries, and provides a summary of larval developmental patterns and behaviors and their potential impacts on larval dispersal and lobster demographic connectivity.

In: The Zoological Guide to Crustacea ISBN: 978-1-53616-366-7
Editor: Noelle Lachance © 2019 Nova Science Publishers, Inc.

Chapter 1

CAVE CRUSTACEAN DECAPODS FROM MEXICO

Luis M. Mejía-Ortíz[*]

Biospeleology and Carcinology Lab., Sustainable Development Division, Quintana Roo University, Quintana Roo, Mexico

ABSTRACT

This chapter is an update of the cave crustacean decapods from Mexico, because in the last thirty years several species have been described. Mexico has diverse cave environments according to the cave origins. There are karstic caves, volcano caves with freshwater habitats and anchialine caves with marine and brackish habitats. The decapods are a conspicuous taxonomic group, and in the caves they have colonized these environments. In Mexico, the following cave species have been exclusively registered; Procarididae (1), Anchialocarididae (1), Atyidae (4) Agostocarididae (2), Alpheidae (3) Barburiidae (1), Hippolytidae (3) Palaemonidae (10), Cambaridae (5), Pseudothelphusidae (5) and Trichodactylidae (4). This chapter include taxonomic keys and a compilation of brief descriptions of each species and their notes on the habitats and distribution on the karstic regions in Mexico.

---

[*] Corresponding Author's E-mail: luismejia@uqroo.edu.mx.

INTRODUCTION

The crustaceans in the caves are an extremely diverse group, because we can find representatives of terrestrial environments (isopods), but also different semi-terrestrial features (crabs) and exclusively in aquatic environments (shrimp, amphipods, among others). The ability presented by crustaceans to inhabit different environments, has meant that even in caves, this group is well represented and is considered among aquatic environments as an important group in the different trophic levels (Mejía-Ortíz, 2019; Alvarez & Illife, 2008).

On the other hand, caves and cenotes are well represented in Mexico and more than 20% of the territory has been reported as having potential for the development of these environments (Espinasa-Pereña 1994; Mejía-Ortíz, 2007), which has offered these crustaceans a variety of places to inhabit. This has led some organisms to enter the cave on an occasional basis (for food, shelter or protection). These animals are called stygoxenes, and there are others that are commonly found in underground environments and they are known as stygophiles. Those that present their entire life cycle in the underground environments coupled with the different adaptations they have to survive in, whether freshwater or marine are known as stygobionts (Camacho, 1992). This work aims to show an update of the decapod crustaceans that live in the caves permanently in Mexico, which has both freshwater and marine representatives and includes eleven families up to now.

CAVES, GROOTES AND CENOTES IN MEXICO: A REGIONAL BACKGROUND

In Mexico there are seven karst provinces according to their features such as origin, soil conformation, physiography and geology, and even palaeography. Inside these provinces, all this diversity offers an excellent opportunity to find a habitat without competition. The seven karst provinces identified are 1) Sierra Madre Occidental; 2) Edwards Plateu and Basins and Rangers; 3) Sierra Madre Oriental; 4) Neovolcanic Plateau; 5) Sierra Madre del Sur Systems; 6) Chiapas –Guatemala Highlands and 7) Yucatán Peninsula (Mejía-Ortíz 2005).

Sierra Madre Occidental

"The northern portion of the Sierra Madre Occidental consists of a series of north-south ranges lying between the ranges of the Sonora desert and the main plateau of the Sierra. It extends from the United States into north eastern Sonora. The principal mass is formed of Tertiary volcanics extending from northern Chihuahua into northern Jalisco where it is bounded on the south by the Río Santiago. The eastern slopes of the Sierra tend to be gentle, but the western side is abrupt and cut by deep gorges, including Barranca del Cobre, with a depth of more than 2,000 meters. Although most of the area is formed of igneous rock, a few isolated outcrops of Cretaceous limestone have been exposed by erosion of the overlying Tertiary deposits" (Reddell, 1981).

Edwards Plateau and Basins and Ranges

"This consists of a narrow band of Cretaceous limestone extending from a few kilometers northwest of Ciudad Acuña to the Serranías del Burro. It has been isolated from the Edwards Plateau proper within the U.S.A. by the incision of the Río Grande.

The Basins and Ranges province consists of an area of folded and faulted mountain ranges separated by wide valleys and basins. The mountains are generally oriented along a Northwest--Southeast axis and frequently enclose large closed basins (bolsones). The most notable of these is the Bolsón de Mapimí in Durango, but the Bolsón de Cuatro Ciénegas de Carranza in Coahuila is of particular importance because of its rich endemic subterranean aquatic fauna. The western ranges are largely igneous, but to the east they tend to be composed of folded Cretaceous sediments. This province extends from the south-western United States into northern Durango and southern Coahuila. It is bounded on the west by the Sierra Madre Occidental, on the eastern by the Sierra Madre Oriental, and on the south by the Cross Ranges of the Sierra Madre Oriental. The average elevation of the filled basins is about 1,200 meters, with the mountains rising up to several thousand meters above the valleys" (Reddell, 1981).

Sierra Madre Oriental

"The Sierra Madre Oriental is a series of folded ranges extending from the Big Bend region of Texas Southeast to Monterrey; here it turns more to the south and extends to Tamazunchale where it turns more to the east to terminate in the Neovolcanic Plateau near Jalapa, Veracruz. Although some igneous rocks occur in the northern portion of the Sierra, the principal rocks are Cretaceous limestone. Jurassic gypsum occurs in valleys between some ranges, especially between Monterrey and Ciudad Valles. The massive Cretaceous limestone forming the Sierra Madre Oriental in this region is highly cavernous, and in many areas there has been extensive surface karst development" (Reddell, 1981).

Neovolcanic Plateau

"The Neovolcanic Plateau forms a great band across central Mexico from the Pacific coast in Jalisco to the Gulf of Mexico east of Jalapa, Veracruz. Hundreds of volcanoes, most now extinct, rise from a comparatively level plain formed by ash and lava deposit. Five volcanoes, Jorullo, Paricutín, Chichonal, Popocatepetl and Colima volcano have erupted in recent times and an eruption south of Mexico City covered prehistoric settlements in the Pedregal de San Angel. A chain of large peaks, including Tancíntaro, Toluca, Popocatepetl, Malinche, Orizaba, and

Cofre de Perote, dominates the landscape. The average elevation in this province is above 2,500 meters, with Orizaba rising to 5,747 meters above sea level. In a few areas erosion has exposed older sedimentary rocks, some of which are know to be cavernous" (Reddell, 1981).

Sierra Madre del Sur System

"According to Raisz (1964), the Sierra Madre del Sur System includes all of the area between the Neovolcanic Plateau and the Isthmus of Tehuantepec, with the exception of a narrow strip along the Pacific Ocean and the Gulf Coastal Lowland to the north. He subdivides this province into five sections: The Balsas-Mexcala Basin, the Oaxaca Upland, the Northeast Folded Ranges, the Southern Slope, and the Northern Section.

The Southern Slope is a mountainous area along the Pacific coast; the Northern Section is the coastal area in Colima and immediate vicinity.

The Balsas-Mexcala Basin is a structurally complex region largely drained by the Tepalcatepec, Balsas and Mexcala rivers. It is bounded on the north by the Neovolcanic Plateau, on the Northwest by the Northern Section, on the south by the Southern Slope and on the east by the Oaxaca Upland. The rivers flow at elevations of 300 to 600 meters, while to the south the land rises to the heights of the Sierra Madre del Sur at 2000 to 3000 meters. To the east the basin widens to form a broad plateau dissected by several rivers. Part of the area is covered by Tertiary volcanics, but Cretaceous limestone is exposed in many places. Along the northern edge of the region, erosion has removed the overlying volcanic rocks to expose heavily karst Cretaceous deposits.

The Oaxaca Upland is a higher area than the basin to the west and has an overall plateau-like character. To the north it is much dissected and in the centre lies the Valley of Oaxaca. The region is geologically complex with granites, metamorphic rocks, and sedimentary deposits exposed. Cretaceous limestone outcrops particularly in the southern part of the upland.

The Northeast Folded Ranges is a disjunctive part of the Sierra Madre Oriental extending from Córdoba, Veracruz to Tehuantepec, Oaxaca. It consists largely of Cretaceous limestone, but volcanic deposit in some areas cover the limestone" (Reddell, 1981).

Chiapas-Guatemala Highlands

"The Chiapas-Guatemala Highlands are a series of arcuate ranges and depressions which begin in Chiapas, and extend generally Northwest-Southeast into Guatemala where they continue more nearly west-east. The first major range is a series of igneous mountains, the northern part of which is the Sierra Madre de Chiapas, rising from the Pacific coastal plain. To the north of this is a major depression known as the Chiapa Depression in Mexico and the Motagua Valley in Guatemala. Rising from this is a second major series of mountains. The southern part of the range is a high, plateau like region that begins as the Sierra de San Cristóbal in Chiapas, becomes the Altos Cuchumates and highlands of Alta Verapaz in Guatemala, and finally the Sierra de Santa Cruz near the Caribbean coast. Along the northern edge, beginning in Mexico as the Sierra de los Lacandones, the ranges descend to form low, knob-like mountains in the southern Petén of Guatemala, and then rise to form the Maya Mountains of Belize.

The Sierra Madre de Chiapas is a batholith of Palaeozoic age composed mainly of granite and diorite partly covered in the southern part by Cretaceous rocks. The Guatemalan continuation is largely mantled by lava and ash deposits. Most of the area is, of course, concavernous, but only two caves in Monzintla region have been studied.

The Chiapa Depression is a wide basin drained by the Río Grijalva and its tributaries. The basin rises gradually from about 700 meters along the Guatemalan border to about 900 meters in the Northwest. The Río Grijalva cuts through the basin, its canyon gradually increasing in depth. It finally dissects the Sierra de San Cristóbal in the great gorge known as El Sumidero. The principal rock units in the area are of Cretaceous age, but some Eocene and Oligocene deposits occur. The Guatemalan portion of this depression is drained by the Río Motagua and is generally more dissected than in Chiapas. Numerous caves are known in the Chiapa Depression, mostly in the vicinities of Tuxtla Gutiérrez, Ocozocoautla, and Malpaso.

The Central plateau of Chiapas is one of the great karst regions in Mexico. The plateau rises steeply from the Chiapa Depression to an elevation of about 2,300 meters. Parts of the plateau, such as near San Cristóbal de las Casas, are rolling plains with numerous shallow dolinas. In other areas there is the greatest development of karst in Mexico, with sinkholes, dolinas, poljes, and sinking streams occurring everywhere on the Cretaceous limestone surface.

The northern side of the plateau is composed of a series of folded ranges in Cretaceous and Tertiary rocks" (Reddell, 1981).

Yucatan Peninsula

"The Yucatán Peninsula is a limestone platform projecting northward from Central America. Northern Yucatán is generally low, with elevations gradually increasing from sea level to about 30 meters in the south. The Sierra de Ticul rises abruptly from the south-western Yucatán plains to elevations of 70 to 100 meters. The Sierra de Bolonchén lies south of the Sierra de Ticual in the state of Campeche. The northern part of the Sierra de Bolonchén is a distinct range, but to the south it is broken into numerous low hills separated by valleys, some containing ponds and lakes. This hilly region continues south into the Petén of Guatemala.

Pleistocene and Holocene deposits outcrop in eastern Campeche and along the northern coast of Yucatán, but most of the Peninsula is covered by rocks of Palaeocene to Pliocene age. The Pliocene and Miocene formations generally outcrop along the coast of the Peninsula with Eocene and older rocks inland. The Sierra de Ticul and most of the hill district of Campeche are formed of rocks of Eocene or Palaeocene age" (Reddell, 1981).

SYSTEMATICS

In Mexico there are 40 cave decapod species involved in eleven families (Figure 1); all around these karstic provinces, and in Yucatan Peninsula there are at least 50% of representative cave fauna, such as we can see in Table 1.

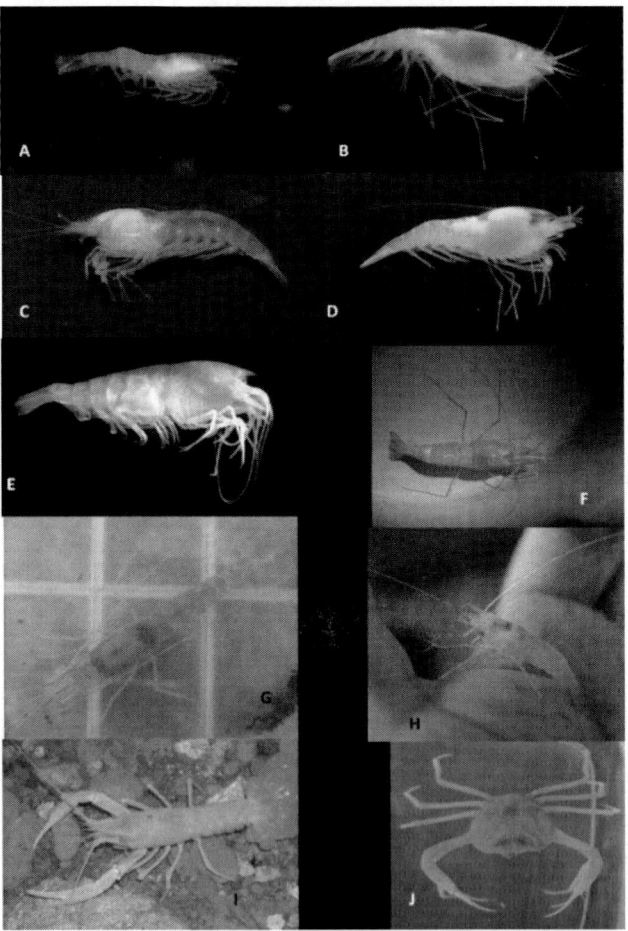

Figure 1. Biodiversity from cave decapods in Mexico: a) Procarididae; b) Anchialocaridae; c) Atyidae (Picture from J. Pakes); d) Agostocarididae; e) Alpheidae (Taken from: tamug.edu); f) Barbouriidae; g) Hippolytidae; h) Palaemonidae; i) Cambaridae; j) Pseudothelphusidae. The remaining pictures are from author.

Table 1. Checklist of cave decapod species and the karstic province in which they are found

Family	Species	Karstic Province
Procarididae	*Procaris mexicana*	Yucatan Peninsula
Anchialocarididae	*Anchialocaris paulini*	Yucatan Peninsula
Atyidae	*Typhlatya campechae*	Yucatan Peninsula
	Typhlatya pearsei	Yucatan Peninsula
	Typhlatya mitchelli	Yucatan Peninsula
	Typhlatya dzilamensis	Yucatán Peninsula
Agostocarididae	*Agostocaris bozanici*	Yucatán Peninsula
	Agostocaris zabaletai	Yucatán Peninsula
Alpheidae	*Yagerocaris cozumel*	Yucatán Peninsula
	Potamalpheops stygicola	Sierra Madre del Sur Systems
	Triacanthoneus akumalensis	Yucatán Peninsula
Barbouriidae	*Barbouria yanezi*	Yucatán Peninsula
Hippolytidae	*Calliasmata nohochi*	Yucatán Peninsula
	Janicea antiguensis	Yucatán Peninsula
	Parhippolyte sterreri	Yucatán Peninsula
Palaemonidae	*Creaseria morleyi*	Yucatán Peninsula
	Neopalemon nahuatlus	Sierra Madre del Sur Systems
	Macrobrachium villalobosi	Sierra Madre del SurSystems
	Macrobrachium acherontium	Chiapas-Guatemla Highlands
	Macrobrachium sbordonii	Chiapas-Guatemla Highlands
	Troglomexicanus perezfarfantae	Sierra Madre Oriental
	Troglomexicanus tamaulipasensis	Sierra Madre Oriental
	Troglomexicanus huastecae	Sierra Madre Oriental
	Cryphiops sbordonii	Chiapas-Guatemla Highlands
	Cryphiops luscus	Chiapas-Guatemla Highlands
Cambaridae	*Procambarus xilitlae*	Sierra Madre Oriental
	Procambarus cavernicola	Sierra Madre del Sur Systems
	Procambarus rodriguezi	Sierra Madre del Sur Systems
	Procambarus oaxacae	Sierra Madre del Sur Systems
	Procambarus reddelli	Sierra Madre del Sur Systems
Pseudothelphusidae	*Villalobosus lopezformenti*	Sierra Madre del Sur Systems
	Typhlopseudothelphusa hyba	Chiapas-Guatemla Highlands
	Typhlopseudothelphusa mocinoi	Chiapas-Guatemla Highlands
	Potamocarcinus leptomelus	Chiapas-Guatemla Highlands
	Odontothelphusa monodontis	Chiapas-Guatemla Highlands
	Pseudothelphusa sonorae	Sierra Madre Occidental
Trichodactylidae	*Rodriguezia mensabak*	Chiapas-Guatemla Highlands
	Rodriguezia villalobosi	Chiapas-Guatemla Highlands
	Rodriguezia adani	Chiapas-Guatemla Highlands
	Avotrichodactylus bidens	Chiapas-Guatemla Highlands

Taxonomic Keys

1	Body compressed; rostrum often with dorsal margin serrate; first abdominal segment not much smaller than others; if first and second pereiopods unequal in size, second larger or carpus of second subdivided; pereiopods often with exopods, fifth not conspicuously smaller than fourth; pleopods natatory, some usually provided with appendices internae, first never modified for sperm transfer	2
1'	Body depressed or not strongly compressed; rostrum, if distinct, always flattened dorso-ventrally and often with lateral (marginal) spines; first abdominal segment always much smaller than others; first pereiopod always larger than second, carpus of second never subdivided; pereiopods never with exopods, fifth occasionally conspicu- ously smaller than fourth; pleopods never natatory nor provided with appendices internae; first or first and second pleopods of male modified for sperm transfer or fifth pereiopod much smaller than fourth	11
2	All pereiopods without chelae	Procarididae
2'	The first and second pereiopods with chelae	3
3	The first and second pereiopod with a rudimentary chelae	Anchialocarididae
3'	The first and second pereiopod with a very well defined chelae, with tufts of setae, or very well chitinous appendage.	4
4	Fingers of chelae of first and second pereiopods with apical brushes of long setae; some pereiopods with exopods	Atyidae
4'	Fingers of chelae of first and second pereiopods without apical brushes of long setae; all pereiopods lacking exopods	5
5	Pereiopod 1, chelate, dactylus more slender than propodal finger, flattened-ovate in cross section, entire cutting edge bearing single row of minute, finely serrate, triangular spines; propodal fixed finger with broad transparent flange along cutting margin, latter having single row of minute, closely set blunt spines; propodal palm about half length of fingers; carpus distally widened, articulating on posterior surface of propodus	Agostocarididae
5'	Articulation between palm from 1st pereiopod with carpus simple on the distal margin	6
6	Carpus of second pereiopod multiarticulate, consisting of 5 or more articles	7
6'	Carpus of second pereiopod undivided	9
7'	Carpus of second pereiopod consisting of fewer than 20 articles; dactyli of third, fourth, and fifth pereiopods without spines on flexor margin; mandible with incisor process	Alpheidae
7'	Carpus of second pereiopod consisting of at least 25 articles; dactyl of third, fourth, and fifth pereiopods with spines on flexor margin; mandible without incisor process	8

8	Rostrum armed with 4 to 6 dorsal teeth and 2 to 4 ventral teeth; mandible with palp	Barbouriidae
8'	Rostrum unarmed; mandible without palp	Hippolytidae
9	Fingers of chelae of first and second pereiopods without apical brushes of long setae; all pereiopods lacking exopods	Palaemonidae
9'	Second and third pereiopod chelate; first pleopod terminating in 2 or more elements; if only 2. cephalic surface with strong	Cambaridae
11	Eye without pigment or faceted cornea; carapace rounded anterolaterally; dactyl of second through fifth pereiopods armed with rows of blunt spines	Pseudothelphusidae
11'	Carapace suborbicular, with 0-5 lateral teeth; front bilobed; postgastric pits absent, abdominarl segments with all sutures visible, or partially visible; third maxilliped with merus trapezoidal, not conspicuously narrow	Trichodactylidae

Species Diagnosis

Procarididae. Chace & Manning 1972. *Procaris mexicana*. Stenberg & Schotte, 2004.

"Integument firm and thin. Rostrum acutely triangular and lacking teeth, only reaching medial concavity of eyes. Carapace devoid of spines; anterior margin distinctly convex and slightly emarginated below distinct cervical sulcus; prominent anteroventral sulcus positioned parallel to ventral margin, and meeting ventral end of cervical sulcus; posterodorsal margin markedly concave. Eyestalk produced into two lobes, the medial lobe sharply triangular and extending beyond the more bluntly triangular lateral lobe; eye lacking facets and with irregular mass of pigment. Antennular peduncle does not reach distal one-third of antennal scale, broad; stylocerite tapering distally to acute apex, almost reaching distal margin of second antennular article; segments subequal in length; anterior margin of basal article with distinct V-shaped dorsomedial cleft. Antennal scale lacking distolateral tooth, distal margin convex, length approximately 2.5 times the width, distal margin of scale reached by antennal peduncle. Pereiopods 1-5 similar in organization, flexor margins lined with simple setae; dactyli approximately 0.12-0.13 times length of propodi, with strong, curved spines. All five pereiopod pairs with straplike exopod; pereiopods 1-4 with distinct simple epipod, and pleurobranch and setobranch; pereiopod 5 lacking epipods, pleurobranch, and setobranch. Third abdominal somite with dorsal cap not reaching middle of fourth somite; posteroventral margin of the six anterior somites broadly rounded. Abdominal sternites 1-5 with median tubercle between coxae of pleopods; sternite 6 with bulbous tubercle posteriorly directed between uropod bases. Telson approximately 1.4 times length of somite 6, not including posterior spines, armed with two pairs of dorsal spines; porterior margin armed with four pairs of spines, lateral spines shortest two mesial pairs roughly half the length of sublateral spines. All pleopods similar in organization, endopods short and weakly development; appendices internae and masculinae absent from all pleopodos" (Stenberg & Schotte, 2004).

Habitat type: Exclusively anchialine systems. Distributional records: The species has been reported only in several cenotes in Cozumel Island: Cenote Tres Potrillos, Cueva la Quebrada, Cenote Aerolito.

Anchialocarididae. Mejia-Ortíz, Yañez & López-Mejía 2017. *Anchialocaris paulini*. Mejia-Ortíz, Yañez & López-Mejía 2017.

"Carapace smooth, without antennal or branchiostegal spines; antennal angle slightly produced anteriorly; pterygostomian angle smoothly rounded, not produced. Carapace with a distinct groove from the suborbital region to the middle of carapace reaching the dorsal area in the middle. Pleura of six abdominal somites semi squared the first segment covered by carapace; posteroventral angles of fourth and fifth subacute. Rostrum dorsoventrally compressed (triangular in cross section), barely overreaching base of first segment of antennular peduncle, without dentition on dorsal and ventral margins. Pleurobranchs on all pereiopods, three arthrobranchs on the third maxilliped, two on first and second pereiopods, one on third and fourth pereiopods, and fifth without arthrobranch. Telson 3.3 times as long as basal width; dorsal spines equidistant from one another. Four pairs of dorsal spines, posterior margin with 1 pair of movable spines and 3 pairs of setae. Lateral ramus of uropods with acute lateral angle, without spines. Eyes acute without pigment, directed anterodorsally. Antennal scale 2.11 times as long as wide; outer margin with distolateral tooth; scale extending beyond distal end of antennular peduncle; flagellum 2 times as long as carapace length.

Antennular peduncle with acute stylocerite reaching distal end of basal segment. First pereiopods with carpus articulating in normal way with propodus, inner margin of fixed finger of propodus serrate and blade-like, distal part of dactylus slender; basis with exopod, reaching the merus; one anterior and two posterior arthrobranchs present. Second pereiopod longer than first pereiopod, weakly chelate; fixed finger slightly shorter than flaccid dactylus, with setae along margin; exopod overpassing the ischium length. Pereiopod 3 longer than pereiopods 1 or 2; dactyl in the inner margin serrate, scattered setae on propodus, carpus and merus, ischium armed with 2 spines; propodus and carpus equal in length, but 0.6 times smaller than merus, and slightly smaller than ischium, arthrobranchs present. Pereiopod 4, propodus with setae on the inner margin; propodus, merus and ischium almost equal in length but longer than carpus; ischium armed with 2 spines; arthrobranchs present. Pereiopod 5, dactyl with seta on the dorsal margin, propodus with setae on the inner margin, 1.7 times longer than carpus, merus and ischium, ischium armed with 2 spines, without arthrobranchs. Endopod of first pleopod of female less than one third length of exopod. Endopod of second pleopod of female with appendix interna more than one-third length of endopod" (Mejía-Ortíz et al., 2017).

Habitat type: Exclusively anchialine systems. Distributional records: This shrimp has been reported only in the locality type The Cenote Chempita in Cozumel Island.

Atyidae de Haan. 1849. *Typhlatya campechae*. Hobbs & Hobbs, 1976.

"Carapace lacking spines. Rostrum triangular, extending anteriorly beyond eyes, and sometimes almost reaching distal end of proximal podomere of antennular peduncle. Anterior margin of carapace produced in broadly rounded antennal lobe and much less prominent pterygostomian lobe. Posterodorsal surface of carapace marked by faint premarginal transverse suture, lateral surface bearing conspicuous hepatic-branchiocardiac groove extending almost entire length of carapace. Pleura of first 3 abdominal somites rounded, those of fourth subacute posteroventrally, and those of fifth virtually acute. Median portion of first abdominal sternite only slightly produced anteroventrally. Sixth abdominal somite approximately twice as long as fifth and with lobe at lateral base of telson rounded; posteroventral margin of sternum produced in short projection in form of isosceles triangle. Telson subequal in length to sixth somite, provided with 2 pairs of spiniform setae dorsally, anterior pair situated slightly ante- rior to base of posterior third, and posterior pair midway between anterior pair and posterolateral extremity; posterior margin rounded and bearing 2 pairs of smooth spiniform setae laterally (lateral pair only about one-fourth as long as mesial pair) and 4, occasionally 5, equally spaced, plumose, spiniform setae between long smooth setae; plumose setae highly variable in length, extending poste- riorly as far as, exceeding, or not reaching level of distal extremity of longest pair of smooth setae. Eye globular, proximomesial portion covered by rostrum but more distolateral portion clearly visible in dorsal aspect; facets and pigment lacking. Antennule with peduncle reaching beyond lateral spine on antennal scale, often to distal margin of latter. Stylocerite acute, falling short of, or slightly, if at all, exceeding distolateral margin of proximal segment of peduncle. Distal podomere of peduncle with dorsodistal plate bearing about 10 setae. Antennular flagella sub equal in length and slightly more than 3 times as long as carapace; lateral ramus with 11 to 13 proximal articles broader than more distal ones; ventral surface of second, third, or fourth through tenth articles each with row of 2 to 5 lanceolate setae borne on distal margin and also often at midlength on all but proximal 2 or 3. Antennal scale approximately 2.3 times as long as broad, lateral border provided with small spine at base of distal fourth. All pereiopods provided with exopods; that of fifth highly variable in degree of development, ranging from exceedingly small and scarcely surpassing midlength of ischium to moderately robust and almost reaching distal end of merus. First pereiopod attaining proximal fourth of antennal scale; carpus, including distal lobe, sub- equal in length to combined length of ischium and merus, and slightly longer than chela. Second pereiopod reaching midlength of an- tennal scale; carpus 1.2 times as long as either chela or merus. Third pereiopod overreaching antennal scale by length of dactyl and one-fourth of propodus; dactyl with 8 spiniform setae on flexor margin; propodus 4 times as long as dactyl and about 1.8 times length of carpus; merus 0.8 to subequal in length to propodus and bearing 3 spiniform setae laterally; ischium subequal in length to dactyl and armed with 1 spiniform seta. Fourth pereiopod overreaching antennal scale by no more, usually less, than length of dactyl, latter bearing 8 spiniform setae on flexor margin; propodus about twice length of dactyl and about 1.8 times length of carpus; merus almost twice as long as carpus, about 1.2 times length of propodus, and bearing 2 submarginal spiniform setae on disto- lateral half; carpus without lateral spine; ischium distinctly longer than dactyl with 1 lateral spiniform seta slightly distal to midlength. Fifth pereiopod extending anteriorly to about same level as fourth; dactyl subequal in length to ischium, flexor margin bearing 45 to 50 denticulate spiniform setae; propodus slightly less than twice length of dactyl and about twice as long as carpus; merus about 0.8 length of propodus and bearing 2 spiniform setae laterally, one slightly distal to midlength, and other near base of distal fifth of podomere; ischium slightly more than half length of merus and lacking spiniform setae laterally. Endopod of first pleopod of male with conspicuous marginal setae, about one-third as long as exopod, and about half length of protopodite. Second pleopod of male with appendix masculina much longer than appendix interna and bearing about 12 apical and subapical spines. Length of endopod of first pleopod of female about two-thirds that of exopod and approximately two-thirds that of protopodite. Length of endopod of second pleopod of female about three-fourths that of exopod and subequal to that of protopodite. Lateral ramus of uropod with small, although prominent, movable spiniform seta laterally at about base of distal fifth. Branchial series composed of 5 pleurobranchs on pereiopodial somites, arthrobranch at base of third maxilliped, and podobranch on second maxilliped; epipods on third maxilliped and on all pereiopods except fifth; and coxal setae on all pereiopods" (Hobbs et al., 1977).

Habitat: Caves with freshwater ponds. Distributional records: Grutas de Xtacumbilxunam, Bolonchenticul, Campeche, Mexico.

Typhlatya mitchelli. Hobbs & Hobbs, 1976.

"Carapace lacking spines. Rostrum subspiniform, not extending anteriorly beyond eyes. Anterior margin of carapace produced in distinct rounded antennal lobe; pterygostomian lobe not well defined. Faint, pre marginal transverse suture on posterodorsal region of carapace. Hepatic branchiocardiac groove prominent and extending almost entire length of carapace. Pleura of first five abdominal somites rounded, that of fifth broadly so and not at all suggestive of being angular. Median portion of first abdominal sternite only slightly produced antero ventrally. Sixth abdominal somite approximately twice as long as fifth, and with rounded lobe at lateral base of telson; posteroventral margin of sternum produced in form of slender triangle. Telson about 0.8 as long as sixth abdominal segment; 2 pairs of spiniform setae ("spines" of most authors) present dorsally, anterior pair situated at base of posterior third and posterior pair approximately midway between anterior pair and mid-posterior margin; posterior margin broadly rounded and bearing short pair of smooth lateral spiniform setae flanked mesially by longer (3.7 times longer than lateral pair) pair of spiniform setae bearing barbules proximomesially; between latter setae, 2 pairs of plumose setae present; 2 mesial pairs of setae somewhat equally spaced but of variable length, either pair being longer or shorter than longer lateral pair. Eye subconical, proximomesial part covered by basal portion of rostrum but more than distal half clearly evident in dorsal aspect; facets and pigment lacking. Antennule with peduncle reaching base of distal third or fourth of antennal scale, almost or quite approximating level of lateral spine of latter. Stylocerite acute, its apex almost reaching or slightly exceeding distal extremity of proximal podomere of peduncle. Distal segment of peduncle with dorsodistal plate bearing 18 to 20 setae. Antennular flagella subequal in length and at least 3 times as long as carapace, sometimes almost as much as 4 times as long; lateral ramus with proximal 15 to 17 articles broader than more distal ones, and ventral surface of fourth, fifth, or sixth through tenth to sixteenth articles with distal pair of lanceolate setae; additional single or paired setae occasionally present at midlength of eleventh through thirteenth articles. Antennal scale approximately 2.5 times longer than greatest width, lateral margin provided with small spine at about base of distal fifth. All pereiopods with exopods, that of fifth reduced, varying in length from barely exceeding distal extremity of basis to reaching distal end of basal fifth of merus. First pereiopod noting (if chela robust) or reaching (if chela slender) almost to distal end of antennal scale; carpus, including distal lobe, subequal to combined length of ischium and merus and 1.1 times as long as chela. Second pereiopod highly variable both in length and in relative degree of development of chela. In individuals with comparatively slender cheliped, chela reaching slightly beyond distal extremity of antennal scale; in those with comparatively heavy cheliped, chela reaching, at most, base of antennal scale; carpus in holotype divided with distal part slightly more than twice as long as proximal one, and combined podomeres twice as long as chela and aboutfive-sixthscombined length of ischium and merus; carpus in animals with slen- der chelae also about twice as long as chela, that in individuals with robust chelae often subequal in length to chela. Third pereiopod overreaching antennal scale by dactyl and three-fourths length of propodus; dactyl with 5 spiniform setae on flexor surface; propodus about 2.7 times longer than dactyl and about 1.5 times longer than carpus; ischium-merus slightly greater than 1.5 times as long as propodus, and bearing 5 submarginal spiniform setae. Fourth pereiopod overreaching antennal scale by half to total length of dactyl; dactyl with 3 spiniform setae on flexor margin; propodus about 2.7 times length of dactyl and about 1.5 times as long as carpus; ischium-merus subequal in length to carpus and propodus combined and bearing 4 somewhat evenly spaced submarginal spiniform setae; carpus with 1 such seta near distal end. Fifth pereiopod overreaching antennal scale by half length of dactyl; dactyl about 2.5 times as long as ischium, flexor margin bearing approximately 40 denticulate spiniform setae; propodus about twice as long as dactyl, 1.8 times as long as carpus, and bearing 2 submarginal spiniform setae near midlength, 1 subdistal and 2 distal ones; merus, bearing 1 submarginal spiniform seta distal to midlength, 1.3 times as long as propodus; ischium about 0.2 length of merus and lacking spiniform setae laterally. Endopod of first pleopod of female small, length 0.3 that of exopod and slightly less than 0.5 that of protopodite. Second pleopod of female with exopod only slightly longer than endopod and about 1.5 times as long as protopodite. Lateral ramus of uropod with prominent movable lateral spiniform setae situated immediately mesial to very small fixed spiniform projection. Branchial series composed of 5 pleurobranchs on pereiopodial somites, arthrobranch at base of third maxilliped, and podobranch on second maxilliped; epipods present on third maxilliped and all but fifth pereiopods, and coxal setae borne on all pereiopods" (Hobbs, et al., 1977).

Habitat: Ponds with freshwater around the Yucatán Peninsula. Distributional records: This species has been reported in several caves and cenotes around the Yucatan & Quintana Roo states (Hobbs & Hobbs, 1976; Hobbs et al., 1977; Alvarez et al., 2015).

Typhlatya dzilamensis. Alvarez, Illife & Villalobos 2005.

"Rostrum unarmed, anteriorly oriented, reaching distal margin of eyes, triangular in dorsal view. Carapace smooth, devoid of spines; in lateral view, dorsal and ventral margins slightly arched; anterior margin, from antennal angle to pterygostomian angle, straight; antennal and pterygostomian angles simple, not produced; posterior margin laterally expanded, overlapping with first abdominal somite. Eyes reduced, not pigmented. Abdomen smooth. Anterior margin of pleura of first somite under posterior portion of carapace. Second somite with pleura broadly rounded, anterior and posterior margins asymmetrical. Pleura of somites 3–4 with posterolateral margin rounded. Fifth somite with posterolateral margin of pleura acute. Sixth somite the longest, 1.4 times length of fifth, middle portion of posterior margin produced. Telson slightly shorter than uropods; elongated, anterior width 2.8 times posterior width; with 2 pairs of articulated, dorsal spines on posterior third. Posterior margin of telson slightly arched, central portion slightly concave; with 3 pairs of spines: external pair the smallest; second pair the largest, 3 times as long as external pair; internal pair 2 times as long as external pair; 2 pairs of setae on middle portion of margin, external pair longer than internal one. Antennular peduncle shorter than antennal scale. Stylocerite well developed, blade-like, ending in sharp tip anteriorly, reaching distal margin of basal segment. Second antennular segment twice as wide as long, segments of internal flagellum shorter. Antenna with scaphocerite twice as long as broad, anterolateral margin with small spine, not reaching distal margin. First pair of pereiopods with exopod reaching proximal third of propodus, distal portion with long plumose setae; carpus hollowed distally, 2.5 times as long as maximum width, longer than propodus, shorter than ischium and merus combined; chela 2.7 times as long as wide, with dense tuft of setae. Second pair of pereiopods longer, more slender than first pair; exopod reaching distal margin of merus, tip with plumose setae; carpus hollowed distally, 4.1 times as long as maximum width, longer than propodus, 0.7 times length of ischium and merus combined; chela 3.2 times as long as wide, with dense tuft of setae. Third pair of pereiopods slender; ischium and merus fused, slightly arched, not straight, with 4 strong spines along ventral margin; exopod 0.6 times length of ischium and merus combined, tip with plumose setae; carpus approximately one-third length of ischium and merus combined; propodus 0.6 times length of ischium an merus combined, with 7 small spines along ventral margin; dactyl with 7 spines on flexor margin. Fourth pair of pereiopods similar to third pair, differing in having dactyl with 8 spines. Fifth pair of pereiopods slender, ischium and merus fused, slightly arched, not straight, bearing 3 spines on distal third of ventral margin; exopod approximately half length of ischium and merus combined, tip with plumose setae; carpus about one-third length of ischium and merus combined; propodus with scattered setae, 0.7 times length of ischium and merus combined; dactyl ending in sharp tip, with line of dense short setae on flexor margin. Epipodites present from third maxilliped to fourth pereiopod, reduced to small blunt hump on fifth pereiopod. First pair of pleopods in both sexes with endopod reduced, devoid of setae, without appendix interna; pairs 2–5 with appendix interna slender, approximately one-third length of endopod. Second pair in males with appendix masculina thicker, 0.8 times length of appendix interna, with rounded distal end bearing 12 setae. Uropods with both rami of about the same length, reaching beyond tip of telson. Endopod oval-shaped, bordered with long setae except for proximal portions of margin. Exopod with straight external margin ending in movable spine, bordered with long setae from posterolateral spine to mesial margin; diaeresis complete, strongly marked laterally, becoming less distinct mesially" (Alvarez et al., 2005).

Habitat: This species lives in anchialine systems. Distributional records: The species was first reported north of the Yucatan Peninsula, but recently was also reported in the Quintana Roo anchialine systems (Alvarez et al., 2015).

Typhlatya pearsei. Creaser, 1936.

"Carapace lacking spines. Rostrum acuminate, reaching midlength to slightly beyond penultimate podomere of antennular peduncle. Anterior margin of carapace produced in prominent broad to narrow and tapering antennal lobe (lobe never acute), followed ventrally by long shallow excavation obliterating dorsal limit of pterygostomian lobe. Posterodorsal surface of carapace lacking shallow premarginal transverse suture, lateral surface bearing conspicuous, sinuous, hepatic branchiocardiac groove extending almost entire length of carapace. Pleura of first 3 abdominal somites rounded, those of fourth subacute, and those of fifth acute to narrowly rounded. First abdominal sternite somewhat broadly produced anteroventrally. Sixth abdominal somite little less than twice as long as fifth and with rounded lobe at lateral base of telson; posteroventral margin of sternum pro- duced in short equilateral triangle forming preanal plate. Telson subequal in length to sixth abdominal somite and provided with 2 pairs of spiniform setae dorsally; anterior pair situated slightly anterior to base of posterior third, and posterior pair midway between anterior pair and posterolateral extremity of telson (asymmetry and double anterosinistral spines in atypical); posterior margin rounded and bearing 2 pairs of smooth lateral spiniform setae (lateral most pair only about one-fourth as long as more mesial pair) and 2 pairs of subequally spaced plumose setae, their length variable but mesialmost pair frequently longest of 4 pairs. Eyes globular, sometimes with slight tuberculiform prominence anterolaterally; postero mesial portion covered by rostrum but much of eye clearly evident in dorsal aspect; facets and pigment lacking. Antennule with peduncle rarely surpassing level of lateral spine of antennal scale. Stylocerite acute, and usually slightly overreaching distal extremity of proximal podomere. Distal podomere of peduncle with dorsodistal plate bearing about 16 plumose setae. Antennular flagella subequal in length and approximately twice length of carapace; lateral ramus with 12 or 13 proximal articles broader than more distal ones, and ventral surface of third through sixteenth articles bearing transverse rows of 2 to 4 lanceolate setae, all except third and sixteenth with 2 rows at midlength and along distal margin. Antennal scale about twice as long as broad, lateral border provided with spine at approximately two-thirds length from base. All pereiopods provided with exopods; that of fifth very much reduced, barely surpassing distal end of basis. First pereiopod reaching end of proximal third of antennal scale; carpus, including distal lobe, about 0.8 as long as ischiomeral podomere and subequal in length to chela. Second pereiopod reaching base of distal third of antennal scale; carpus almost 1.5 times as long as either chela or merus. Third pereiopod overreaching antennal scale by at least length of dactyl and sometimes by as much as additional one-fourth of propodus; dactyl with 9 spiniform setae on flexor surface; propodus almost 2.5 times as long as dactyl; carpus about half as long as propodus and with single spiniform seta laterodistally; merus approximately 1.2 times length of propodus and bearing 4 spines laterally; ischium about half as long as dactyl and lacking spiniform setae. Fourth pereiopod overreaching antennal scale by half to total length of dactyl; dactyl shorter or subequal in length to ischium, its flexor margin armed with 8 spiniform setae; remaining podomeres with setae as on third. Fifth pereiopod reaching distal end of antennal scale; dactyl about twice as long as ischium, its flexor margin bearing 40 or more denticulate spiniform setae; propodus slightly less than twice length of dactyl, almost twice as long as carpus, and bearing row of 3 marginal slender spiniform setae; merus little less than 1.5 times length of propodus and bearing as many as 3 lateral spiniform setae; ischium about one-sixth as long as merus and lacking spiniform setae. Endopod of first pleopod of male with conspicuous marginal setae and slightly more than one-third as long as exopod and little more than two-thirds length of protopodite. Second pleopod of male with appendix masculina much longer than appendix interna and bearing about 16 apical and subapical spines. Length of endopod of first pleopod of female about one-fourth that of exopod and little less than one-half length of protopodite. Second pleopod of female with endopod 0.8 as long as exopod and about two-thirds as long as protopodite. Lateral ramus of uropod with small movable spiniform seta situated in lateral angular excision at base of distal fifth. Branchial series composed of 5 pleurobranchs on pereiopodial somites, arthrobranch at base of third maxilliped, and podobranch on second maxilliped; epipods on third maxilliped and on all (except fifth) pereiopods; and coxal setae on all pereiopods" (Hobbs et al., 1977).

Habitat: This species has been reported to inhabit the anchialine systems but also they can move through the halocline in the same system and go to the freshwater layer. Distributional records: This species lives in several systems in Yucatan and Quintana Roo states (Hobbs et al., 1977; Holthuis, 1977; Alvarez et al., 2015).

Agostocarididae. Hart & Manning, 1986. *Agostocaris bozanici*. Hart & Manning, 1986.

"Integument firm, thin, with sparse scattering of tiny black chromatophores. Rostrum triangular, apically acute, bilaterally compressed, dorsally unarmed, carinate, carina extending onto anterior carapace for distance equal to rostral length, ventrally carinate. Dorsal profile of carapace evenly convex; infraorbital and branchiostegal angles rounded; very faint orbito branchial sulcus present. Pleuron of abdominal somite 2 broadly ovate; pleura of somites 3-5 posteroventrally rounded; somite 6 dorsally twice length of somite 5, with small posteroventral tooth. Telson about 3 times longer than basal width, tapering slightly, bearing 5 pairs of mobile lateral spines; posterior margin slightly convex, armed with 5 pairs of spines, second pair from lateral margin longest. Eye not differentiated into cornea and stalk, lacking pigment, conical, directed anterodorsally. Antennule with basal peduncular article subequal in length to articles 2 and 3 combined; stylocerite with proximal two-thirds parallel sided, apically acute; superior flagellum subequal to carapace in length, inferior flagellum slightly longer. Antennal scaphocerite 1.7 times as long as greatest width, small distal tooth on lateral margin not reaching apex of blade; flagellum at least 3 times longer than carapace length. Pereiopod 1 shorter than maxilliped 3 and pereiopod 2, chelate, dactylus more slender than propodal finger, flattened ovate in cross section, entire cutting edge bearing single row of minute, finely serrate, triangular spines; propodal fixed finger with broad transparent flange along cutting margin, lat- ter having single row of minute, closely set blunt spines; propodal palm about half length of fingers; carpus distally widened, articulating on posterior surface of propodus. Pereiopod 2 more slender but longer than pereiopod 1, chelate, dactylus lamellate, margins setose; propodal finger heavier than dactylus, faintly curved along outer margin, compressed, with numerous setae. Pereiopod 3, ischium bearing 2 spines on outer surface; dactylus having 3 spines on posterior surface. Pereiopod 4, ischium bearing 2 spines on outer surface; propodus with cluster of serrate grooming setae posterodistally; dactylus having 3 spines on posterior surface. Pereiopod 5, ischium with 2 spines on outer surface; propodus 1.8 times length of propodus of pereiopod 4, bearing dense serrate grooming setae along posterodistal third; dactylus with row of about 8 small spines on posterior margin. Pleopod 1 biramous, exopod elongate, setose; endopod short, ovate, devoid of spines or setae. Pleopods 2-5 biramous, exopod and endopod elongate, setose; endopod bearing slender appendix interna. Outer uropodal ramus bearing 2 spines on outer distal margin, diaeresis in distal third, distal margin broadly rounded; inner uropodal ramus narrower and distally more narrowly rounded than outer ramus" (Hart & Manning, 1986).

Habitat: This shrimp lives in the anchialine systems with a high sulphur content. Distributional records: The species has been reported in only one localty: Cenote Xcan-ha in Cozumel Island.

Agostocaris zabaletai. Mejía-Ortíz, Yañez & López-Mejía, 2017.

"Integument firm, thin. Rostrum triangular, apically acute, bilaterally compressed, dorsally and ventrally unarmed, carinate. Dorsal profile of carapace evenly convex; infraorbital and branchiostegal angles rounded; very faint orbitobranchial sulcus present. Pleuron of abdominal somite 2 broadly ovate; pleura of somites 3-5 posteroventrally rounded; somite 6 dorsally 1.3 times the length of somite 5. Telson about 2.6 times as long as basal width, tapering slightly, bearing 4 pairs of mobile lateral spines; posterior margin slightly convex, armed with 5 pairs of tooth, second pair from lateral margin longest. Outer uropodal ramus bearing 1 spine on outer distal margin, diaeresis in distal third, distal margin broadly rounded; inner uropodal ramus narrower and distally more narrowly rounded than outer ramus. Eyes, lacking pigment, conical, directed anterodorsally. Antennal scaphocerite 2.2 times as long as greatest width, small distal tooth on lateral margin not reaching apex of blade; flagellum at least 2 times as long as carapace length. Antennule with basal peduncular article shorter than articles 2 and 3 combined; stylocerite with proximal two-thirds parallel-sided, apically acute. Pereiopod 1 shorter than maxilliped 3 and pereiopod 2, chelate, dactylus more slender than propodal finger, flattened ovate in cross section, entire cutting edge bearing single row of minute, finely serrate, triangular spines; propodal a fixed finger with broad transparent flange along cutting margin, latter having single row of minute, closely set blunt spines; propodal palm about 0.8 times the length of fingers; carpus distally widened, articulating on posterior surface of propodus, Merus 2.3 times longer than ischium, basis with exopod ending in several setae almost reaching the distal border of merus, arthrobranch present. Pereiopod 2 more slender but longer than pereiopod 1, chelate, dactylus lamellate, with margins setose;

propodal finger heavier than dactylus, faintly curved along outer margin, compressed, with numerous setae, carpus slightly smaller than merus and similar in size than ischium, which have a single seta, basis with a exopod surpassing the distal border of ischium ending in serveral setae arthrobranch and pleurobranch present. Pereiopod 3; dactylus having 5 spines on posterior surface, propodus in the ventral margin with three spines, carpus 1.75 times smaller than merus, ischium bearing 2 spines on outer surface, arthrobranch present. Pereiopod 4; dactylus with 4 spines on posterior surface, propodus with cluster of serrate grooming setae posterodistally; carpus 0.6 times smaller than merus and 0.8 of ischium, ischium bearing 2 spines on outer surface, basis without spines and coxa with arthrobranch. Pereiopod 5, dactylus with row of about 7 small spines on posterior margin; propodus 1.4 times length of propodus of pereiopod 4, bearing dense serrate grooming setae along posterodistal third; carpus and merus equal in length, ischium with 2 spines on outer surface, basis short and coxa with arthrobranch. Pleopod 1 biramous, exopod elongate, setose; endopod short, ovate, devoid of spines or setae. Pleopods 2-5 biramous, exopod and endopod elongate, setose; endopod bearing slender appendix interna. Appendix interna two thirds length of endopod" (Mejia-Ortíz et al., 2017).

Habitat: This species lives in the anchialine systems in Cozumel Island. Distributional records: The species has been recorded on Cenote Tres Potrillos; Sistema Quebrada and Cenote Chempita in Cozumel Island. Alpheidae. Rafinesque, 1815. *Yagerocaris Cozumel.* Kensley, 1988.

"Integument of carapace and terga and pleura of abdomen bearing fairly dense, very short setules. Rostrum broad based, triangular in dorsal view, ventral margin faintly sinuous in lateral view, reaching to about second antennular peduncle article. Carapace margin slightly convex in antennal region; pterygostomian region produced into strong spine. Anterior margin of pleuron 1 straight; pleuron 2 more broadly rounded posteroventrally than anteroventrally; pleura 3-5 each with distinct posteroventral tooth. Abdominal somite 6 four-fifths middorsal length of somite 5, with acutely triangular posterolateral lobe. Telson with 2 pairs of articulated dorsal spines in posterior half; posterior margin bearing 2 pairs of spines, outer pair one-fourth length of inner pair; margin between inner pair of spines produced into rectangular lobe reaching more than halfway along elongate inner spines, with 18-20 plumose setae on truncate/rounded posterior margin. Eyestalks basally fused, anteriorly flattened, small weakly pigmented area flanked by pair of low subacute lobes. Antennule with distally acute stylocerite reaching beyond distal margin of second peduncle article; basal peduncle article subequal in length to 2 distal articles together; dorso lateral flagellum slightly longer than carapace plus rostrum, fused basal portion of 5 articles, free part of shorter ramus single article bearing 3 clumps of aesthetascs; ventromesial flagellum subequal to dorsolateral flagellum in length. Antennal scaphocerite about two-thirds longer than greatest width; lateral margin nearly straight, ending in strong tooth reaching as far forward as anterior margin of blade; flagellum almost 3 times carapace length. Pereiopod 1 more robust but shorter than following legs, chelate, barely reaching base of antennal scale; dactylus about two-thirds length of propodal palm; cutting edges of fingers entire, straight. Pereiopod 2, both members subequal, overreaching maxilliped 3 by chela and at least distalmost carpal article; carpus consisting of 4 distal sub- equal articles plus more elongate proximal article. Pereiopod 3, ischium bearing 2 articulating spines on posterior margin; merus with single articulating spine at about mid-length of posterior margin. Pereiopod 4 sub-equal to pereiopod 3, ischium with 2 spines on posterior margin, merus with 1 or 2 spines on posterior margin. Pereiopod 5 little longer than pereiopod 4, ischium with 1 spine on posterior margin; merus with 1 spine on posterior margin; propodus with postero distal series of overlapping transverse rows of elongate grooming setae. Pleopod 1, endopod about one-fourth length of exopod, with well-spaced marginal setae. Pleopod 2, endopod with appendix masculina and appendix interna articulating at about midlength of mesial margin; appendix interna one-third length of appendix masculina, latter reaching beyond apex of rami, rodlike, with few distal spines. Uropodal rami barely overreaching telsonic apex; lateral ramus with articulating spine mesial to tooth on outer margin just overreaching distal margin of blade; diaresis complete" (Kensley, 1988).

Habitat: This species has been recorded in anchialine systems. Distributional records: The species was reported in Cozumel Island, in

Cenote Aerolito, but it has not been seen by divers in the last 15 years. It has been recently reported in the Yucatan Peninsula (Alvarez et al., 2015).
Potamalpheus stygicola. Hobbs, 1973.

"Rostrum triangular with strongly acute apex not reaching midlength of proximal segment of antennular peduncle, and flanked laterally by pair of acute supraorbital spines. Carapace with pair of incipient carinae diverging posteriorly from lateral bases of supraorbital spines; hepatic region with several complexly arranged grooves, and conspicuous, deep, sclerotized hepatic-branchiocardiac groove extending almost entire length of carapace. Pterygostomial margin rounded, lacking spine; posterior margin with moderately prominent cardiac notch at base of branchiostegite. Setae on the anteromesial surface of the eyes, Movable spines on the merus of one or more of the third through fifth pereiopods, and a toothed lamella along the diaeresis of the lateral ramus of the uropod. Four anterior abdominal somites lacking median carina dorsally, all with rounded pleura; pleuron of fifth abdominal somite with acute posteroventral angle. Sixth somite slightly longer than fifth and about three-fourths as long as telson, its posteroventral angle consisting of acute triangular articulated plate, margin rounded at base of telson. Telson about 2.4 times longer than wide with paired dorsal spines situated at midlength and additional pair at base of distal fourth; rounded posterior margin with two pairs of lateral spines, more mesial pair longer flanking row of 18 plumose setae. Eyes almost covered by carapace, with pigment greatly reduced, fused stalks with antero-median bulge or with paired submedian anterior budges. Antennular peduncle with acute stylocerite almost reaching distal extremity of proximal podomere; second podomere about 1.3 times longer than third; flagella approximately twice length of carapace, lateral flagellum with nine or 10 articles proximal to bifurcation and short branch consisting of only three articles, distal most indistinctly delimited basally. Antenna with peduncle almost reaching base of distal third of scale; proximal segment with small acute distolateral tooth, and second segment with ventrodistal spine; flagellum about three times as long as carapace; antennal scale about 2.2 times longer than broad, with acute distolateral tooth not reaching so far distally as distal margin of blade. First pereiopods reaching slightly beyond midlength of antennal scale, subequal in size, carpus and chela subequal in length and only slightly shorter than merus; opposable margins of fingers of chela without prominent teeth or spines; coxa with epipodite consisting of α and β components. Second pereiopod overreaching antennal scale by slightly more than length of dactyl, with chela similar to that of first; carpus consisting of five articles, proximal most longer than combined length of second, third, and fourth and almost twice as long as distal article; merus distinctly longer than three proximal articles of carpus and also longer than ischium; coxa with epipodite as in first pereiopod. Third pereiopod overreaching antennal scale by length of dactyl and half that of propodus, with simple dactyl; propodus 2.7 times longer than dactyl and 1.2 times longer than carpus; merus slightly longer than propodus; ischium distinctly shorter than carpus; coxa with epipodite as in first and second pereiopods. Fourth pereiopod, overreaching antennal scale by slightly more than length of dactyl, shorter than third; dactyl simple; propodus 2.2 times length of dactyl, 1.4 times than of carpus, and subequal in length to merus; ischium shorter than carpus; coxa with epipodite as in first pereiopods. Fifth pereiopod overreaching antennal scale by about two-thirds length of dactyl, subequal in length to third; dactyl simple, propodus, bearing transverse rows of setae on distal remoter surface, 2.9 times longer than dactyl and 1.1 times longer than merus; ischium much shorter than carpus; coxa with epipodite limited to setiferous (β) element. First pleopod with exopodite 2.8 times longer than endopodite. Second pleopod with exopodite approximately 1.2 times longer than endopodite; appendix masculina naked except for six apical spine like setae. Lateral ramus of uropod with entire, straight lateral margin terminating in short acute tooth, and longer movable spine situated immediately mesial to tooth" (Hobbs et al., 1977).

Habitat: This shrimp lives exclusively in freshwater caves. Distributional records: The species has been registered in the caves from Sierra de San Antonio and Gabriel Cave in the Acatlán region from Oaxaca State (Mejia-Ortíz, 2005).

Triacanthoneus akumalensis. Alvarez, Illife, Gonzalez & Villalobos, 2012.

"Carapace broadly rounded laterally, with faint anterolateral suture and two strong, acute postorbital teeth; dorsal midline with prominent sharp tooth in epigastric position, low carina present between epigastric tooth and orbit; posterior margin with deep cardiac notch. Rostrum reaching mid-length of third antennular article, tip slightly pointing upwards, ventral margin with small tooth. Eyes completely covered by carapace, not visible in dorsal view; lateral region pigmented, cornea not faceted; dorsal margin with small acute projection present. First three abdominal somites with lateroventral margins of pleura rounded; fourth somite with lateroventral margin subtriangular; fifth somite with lateroventral margin produced in a small acute tooth; sixth somite longer than other somites, without articulated plate. Telson approximately 2.6 times as long as its anterior width; posterior margin with deep "V" shaped notch, each posterolateral angle with a pair of long spines, lateral longest; posterior half of dorsal surface with shallow median longitudinal groove and two pairs of short spines. Uropods with both rami longer than telson, rounded distally; exopod longer than endopod, with complete diaeresis and strong distolateral spine. Antennule with prominent stylocerite, its acute tip reaching beyond mid-length of second article; second article longer than wide, about as long as third article; external flagellum with four basal articles, accessory branch with about eight articles bearing long aesthetascs. Antenna with basicerite armed with stout, sharp distal tooth; carpocerite short, not reaching beyond half-length of scaphocerite; scaphocerite ovate, with rounded anterior margin reaching beyond distolateral tooth. First pair of pereopods asymmetrical, major cheliped always on the right side. Major cheliped 0.75 times of total body length; ischium about one third of merus length, with three ventral spines; merus with ventral margin depressed, dorsal margin straight, distally somewhat thickened; carpus shorter than merus, wider distally; propodus strongly arched; palm about 2.4 times as long as high, inflated, with low hump on dorsal surface, adjacent to articulation with dactylus; fingers about 1.3 times length of palm, strongly crossing distally; cutting edge of fingers serrated with minute teeth, and with four-five conspicuously larger teeth intercalated among smaller teeth. Minor cheliped approximately half as long as major cheliped, much less stout; ischium with three ventral spines; ischium, merus, carpus and propodus subequal in length; chela simple, dactylus slightly shorter than palm. Second pair of pereopods slender; ischium and merus subequal in length; ischium with row of three spines on ventrolateral surface; carpus five-articulated, first carpal article longest, more than half-length of the entire carpus, remaining carpal articles of about the same length; chela simple, with fingers longer than palm. Third to fifth pairs of pereopods slender with simple dactyli, increasing in length posteriorly; third and fourth pereopods each with three spines on ventrolateral surface of ischium; carpus longest article in third and fourth pereopods; propodus longest article in fifth pereopod. Pleural process, strong and anteriorly arched, arising from lateral body wall, above coxa of fifth pereopod. Second pair of pleopods of male paratype with protopod bearing long thick setae curling distally; exopod longer than endopod; endopod with appendix interna and appendix masculine, former elongate, slightly longer than appendix masculina; appendix masculina cylindrical, with nine strong spines along lateral and distal margins" (Alvarez et al., 2012).

Habitat: Exclusively in anchialine caves. Distributional records: This species was reported only in a locality in the Akumal area from Quintana Roo State.

Barbouriidae. Christoffersen, 1987. *Barbouria yanezi*. Mejía, López & Zarza 2008.

"Medium sized shrimp, maximum total length 64 mm. Rostrum slightly large, straight, tip almost reaching distal border of second antennular segment; dorsal margin bearing 5 teeth, 3 in postorbital position, 5 teeth on ventral margin. Carapace smooth, maximum length 20.9 mm, with antennal spine equal to banchiostegal spine. Branchiostegal groove shallow. Abdomen smooth, first somite with anterior margin of pleuron straight, second somite with rounded pleura, somites 3-5 with posterior angle of pleura acute, and each with a posterolateral tooth. Posteroventral margin of third and fourth somites bearing setae on ventral border. Sixth somite 2.2 times as long as fifth. Telson 1.4 times longer than sixth somite, almost equal to uropodal rami; bearing three pairs of dorsal spines, first pair in its proximal third, second pair in middle section, and the third pair almost on posterior border of telson; posterior margin rounded, bearing one pair of lateral spines, larger than inner pair; external pair 4.6 times as long as internal pair. Eyes pigmented, cornea narrower than stalk. Antennules with acute stylocerite not reaching distal margin of first peduncular segment. First antennular segment with concave depression to fit eye. Second antennular segment cylindrical. Antennae with basicerite smooth. Scaphocerite 2.47 times as long as wide, distolateral spine short, widely separated from distal margin of main blade. First pereiopods shorter than all other pereiopods, and slightly robust, with tufts of setae on articulation of dactylus and fixed finger. Tip of fingers not surpassing distal margin of scaphocerite; palm slightly compressed, almost as long as dactylus; dactylus curved inwards; carpus 2 times length of palm, 0.7 times length of merus. Second pair of pereiopods subequal in size, without spines. Palm semicylindrical, as long as wide, and with length equal to that of dactylus; carpus 26 times palm length, 2.28 times as long as merus; both multiarticulate, carpus with 23 and merus with 11 segments, therefore ischium also multiarticulate in the distal portion with 4 segments; ischium 1.08 times merus length. Fingers not gaping, cutting margins covered with rows of tufts of setae, both propodus and dactylus with setae on apical area. Propodus and dactylus of third pereiopod sparsely pilose; distal spines on articulation with dactylus. Propodus 6.8 times length of dactylus, 1.17 times carpus length. Two rows of 8 spines on inner and distal border of merus. Fourth pereiopods sparsely pilose; propodus 9 times dactylus length, 1.5 times as long as carpus; with two rows of 6 spines on inner and distal border of merus, four setae on propodus–dactylus articulation. Propodus of fifth pair of pereiopods pilose; two spines on inner border of merus, four setae on propodus–dactylus articulation; propodus 12 times dactylus length, 1.7 times carpus length. Appendix masculina 0.6 times length of appendix interna, 11 spines on its apical portion" (Mejía et al., 2008).

Habitat: Exclusively in anchialine systems. Distributional records: This species has been recorded in two cenotes in Cozumel Island: Cenote Tres Potrillos & Sistema Quebrada.

Hippolytidae. Dana, 1852. *Calliasmata nohochi Escobar-Briones.* Camacho & Alcocer, 1997.

"Integument of carapace finely pitted with minute sharp, elongated scales implanted. Scales more densely implanted on anterior surface of carapace. Rostrum formed by simple spine, directed ventrally in lateral view, reaching distal half of eye peduncle, upper margin concave, lower margin sharply pointed, concave, reaching half distance of antennal spine. Antennal spine reaching distal end of ocular peduncle. Pterygostomial region rounded, not produced anteriorly, posterior margin without cardiac notch. Abdomen smooth, sparsely covered by scales, setules, or pits. Pleura 1-4 broad, sub- quadrate, posterior angles devoid of spines. Pleuron of first somite broad, anterior and posterior angle slightly rounded, devoid of spines; pleuron of second somite wide, rounded with short, blunt lobe on distal margin; pleura 3 and 4 with distal margin rounded, without ventral spine. Anterior margin of pleura 5 and 6 rounded; posteroventral angle of fifth segment acute. Pleuron of sixth somite bearing acutely triangular posterolateral lobe, devoid of movable plate. Somite 6, 1.3 times length of somite 5 and 0.7 middorsal length of telson. Telson broken at distal tip. Upper of telson bearing 2 pairs of dorsal spines, first pair placed on anterior half of telson, second pair closer to posterior margin. Eyestalks free, not fused laterally or dorsally, close together, short, broad, triangular, reaching over end of antennal spine of carapace; distal portion of lateral margin with scalelike projection. Pigment on eyes dispersed as irregular black stains. Antennule with triangular stylocerite with blunt end in lateral view, reaching half of first peduncle article, lateral margin with sparse long setae. First segment of antennular peduncle 1.3 times longer than segment 2 and 2.3 times longer than segment 3; dorsolateral flagellum with 2 rami, fused basal portion consisting of 21-27 segments; shorter free ramus consisting of 1 single reduced article; inner ramus slender, consisting of single, narrow, multiarticulate branch. Scaphocerite well developed, twice as long as wide, reaching distal end of second segment of antennular peduncles, outer margin straight, ending in small tooth, directed anteroventrally, reaching anterior margin of blade. Carpocerite 0.66 length of scaphocerite. In lateral view with 2 spines distally, one above (covered by scaphocerite) and one below at base of scaphocerite. First pereiopods with well-developed chelae, more robust but shorter than in following legs, subequal in size and shape, reaching antennular peduncle; chela ovate, 3.5 times longer than wide, dactylus with sharply pointed tip, dark colored, 0.8 times length of propodal palm, lower margin bearing tufts of long setae, cutting edges with dispersed setae; carpus rectangular, 0.6

times length of merus, 2.2 times longer than wide and 0.6 times length of chela, merus 3.8 times longer than wide, subequal to chela with row of strong spines and long, slender setae on lower margin. Merus 1.2 times longer than chela. Ischium 3.7 times longer than wide, bearing small spines and long setae on lower margin. Second pereiopods long and slender, length unequal. Right longer than left, reaching distal margin of merus end of antennular peduncle. Chelae small, narrow and elongated, 4.5 times longer than wide, finger 0.8 times length of palm, dactylus equa"y long as palm ending in sharply pointed tips bent ventrally, bearing tufts of long setae.Carpus multiarticulated, 34 segments, 10.4 times longer than chela, 1.2 times longer than merus, and 1.3 times longer than ischium. Merus length subequal to ischium with 20 annulations. Ischium divided in 4 segments proximal being longest; inner margin with row of 8 spines with curved tips and several short stiff hairs. Left second pereio pod not overreaching antennal peduncle. Small, narrow, elongated chelae 2.6 times longer than wide, dactylus 0.3 times length of palm, bearing setae. Carpus with 28 segments, 14 times longer than chelae, 1.4 times longer than merus, and 1.8 times longer than ischium. Merus with 16 annulations, subequal in length to ischium. Ischium with segmentation similar to longer pereiopod, inner margin with row of 8 spines with curved tips and several short stiff hairs. Third pereiopod overreaching antennal peduncle by distal margin of carpus. Dactylus slender, 4 times as long as wide, ending in Sharp pointed tip, curved ventrally, and lower margin armed with 4 robust spines; flexor margin defined with line from tip of dactylus; propodus approximately 4.3 times as long as dactylus, with 10 spines combined with setae; carpus 1.5 times longer than propodus, with 8 pairs of spines along posterior margin; merus 8.3 times longer than wide, 1.3 times as long as carpus, armed with 4 strong, robust spines; ischium 3.3 times longer than wide, with strong spine on proximal margin. Fourth pereiopod narrower than third, dactylus similar, more than 4.4 times as long as wide; propodus 5 times as long as dactylus, its upper margin bearing tufts of setae, lower margin provided with 16 pairs of spines with scattered tufts of setae; carpus 15.3 times as long as wide, 1.5 times as long as propodus; carpus with distal projection ending in long distal setae, row of 11 spines along lower margin. Merus 8.7 times longer than wide, 1.3 times as long as carpus. Lower margin of merus and ischium with single spine each.

Ischium 2.8 times longer than wide. Fifth leg narrower than previous 2 legs, extending with propodus beyond antennal peduncle; dactylus slender, 4 times as long as wide, ending in curved tip and bearing 4 posterior spines; propodus 7.1 times as long as wide, posterior margin bearing 16 pairs of small spines, distal portion of margin provided with row of long, dense tufts of hairs and se- tae continuing on anterodistal margin; carpus 0.5 times as long as propodus and 10 times longer than wide, its posterior margin devoid of spines; merus 1.5 times longer than carpus" (Escobar-Briones et al., 1997).

Habitat: This shrimp lives exclusively in anchialine systems. Distributional records: The species was recorded in caves in Quintana Roo continental, but recently was recorded in Cozumel Island from Cenote Tres Potrillos.

Janicea antiguensis. Chace, 1972.

"Carapace with antennal and branchiostegal teeth; rostrum slender, about ve times as long as high, short, extending to about end of basal article of antennular peduncle, with three to four dorsal (one or two postorbital) teeth and one ventral tooth. Eyes large, well-pigmented, cornea broader than eyestalk. Anterior four pleura rounded, fifth pleuron acute posteroventrally. Telson with two pairs of dorsal spines and three pairs of terminal spines, mesial spines longest. Mandible lacking incisor process, with three-jointed palp. First and second pereopods chelate; merus, carpus, and propodus of second pereopod multiarticulate; third to fth pereopods with carpus and propodus multiarticulate. Endopod of male first pleopod without appendix interna, with distal coupling hooks; endopod of male second pleopod with appendix masculina longer than appendix interna" (Giraldes et al, 2012).

Habitat: Exclusively in anchialine systems. Distributional records: It has been recorded in several caves only from Cozumel Island.

Parhippolyte sterreri. Hart & Manning, 1981.

"Rostrum laterally compressed, reaching 2nd joint of antennular peduncle, bearing 3 or 4 teeth on dorsal margin and 4 or 5 on ventral margin; 1 or 2 dorsal teeth distinctly postorbital. Carapace smooth, bearing antennal and branchiostegal spines; antennal angle produced anteriorly beyond base of antennal spine; pterygostomian angle rounded, not produced. Pleura of first 3 abdominal somites broadly rounded; 4th rectangular but unarmed, 5th armed posterolaterally in male, 5th and 6th so armed in female. Telson 5 times as long as broad, tapering in width posteriorly; anterior pair of dorsal spines situated slightly posterior to midlength. Two pairs of movable posterior spines (lateral ones twice length of mesial ones), and posterior margin forming 2 fixed lateral spines and single median fixed spine. Lateral ramus of uropod with acute lateral angle, bearing single artic- ulated spine mesial to angle. Eyes prominent, with black corneas. Antennular peduncle with acute stylocerite not reaching distal end of basal segment. Second segment subequal in length to third. Antennal scale 3 times as long as wide; outer margin straight throughout most of length, ending in tooth; scale extending well past distal end of antennular peduncle. Pereiopods 1 and 2 chelate. Pereiopod 2 longer than 1; carpus and merus subdivided, ischium with several rudimentary subdivisions in distal, 1/5. Pereiopods 3-5 with propodus subdivided, merus armed with conspicuous spines. Endopod of st pleopod of male slightly more than 1/2 length of exopod; without appendix interna. Endopod of 2nd pleopod of male with robust appendix masculina subequal in length to appendix interna and bearing distal ring of setae; longest setae about 1/2 length of appendix masculina" (Hart & Manning, 1981).

Habitat: Exclusively in anchialine systems. Distributional records: It has only been recorded in Cozumel Island.

Palaemonidae. Rafinesque, 1815. *Creaseria morleyi*. Creaser, 1936.

"Eyes reduced and without pigment. Rostrum with dorsal and ventral teeth. Carapace with antennal and branchiostegal spines, lacking hepatic spine. Mandible with 2-jointed palp. Second pereiopod slightly more robust than first, both chelate. Remaining pereiopods with simple dactyl; propodus of fifth pereiopod with numerous trans- verse rows of setae on posterodistal surface" (Hobbs et al., 1977).

Habitat: This shrimp was reported mostly in freshwater caves, but there are some records in anchialine systems too. Distributional records: The species has a wide distribution around the Yucatan Peninsula, in Campeche, Yucatan and Quintana Roo State.

Neopalemon nahuatlus. Hobbs, 1973

"Eyes reduced, without pigment or faceted cornea. Rostrum with teeth dorsally and ventrally. Carapace with antennal and hepatic spines but lacking supraorbital and branchiostegal spines. Telson with 2 pairs each of dorsal and posterior spines and 5 pairs of plumose setae between mesial pair of posterior spines. Mandible with 2- segmented palp. Second maxilliped with epipodite, third with 2 arthrobranchs. Second pereiopod not markedly robust; third through fifth with simple dactyl; propodus of fifth pereiopod lacking trans- verse rows of setae o n posterodistal surface. Appendix interna absent on first pleopod; appendix masculina distinctly overreaching appendix interna" (Hobbs et al., 1977).

Habitat: Freshwater cave. Distributional records: Only reported in a locality type in Oaxaca State.

Macrobrachium villalobosi. Hobbs, 1973.

"Rostrum moderately high and almost straight, only slightly arched immediately anterior to posterior margin of orbit, and its tip reaching slightly beyond distal extremity of antennal scale; dorsal margin with nine to 11 teeth (nine in holotype), one or two, rarely three, (epigastric) of which situated posterior to orbit, and one to three, usually two on ventral margin. Carapace with antennal and hepatic spines, latter situated on level slightly posterior to basal epigastric spine. Branchiocardiac groove prominent. Abdomen smooth; pleura of fourth and fifth somites with angular posteroventral extremities, that of fifth acute. Sixth somite about 1.4 times longer than fifth, and telson with anterior pair of spines situated at base of penultimate fifth and posterior pair nearer posterior margin of telson than to anterior pair of spines; posterior margin of telson contracted to form acute median tip, bearing two pairs of spines arising ventral to margin, more mesial ones extending posteriorly slightly beyond tip of telson (third spine present on left side in holotype) and row of about

10 plumose setae between mesial pair of spines; single pair of fine submarginal setae present dorsally. Eyes moderately large, slightly cleft distally and without facets; distolateral area without trace of pigment or tinted with diffuse reddish purple granules in degenerate corneal area. Antenule with proximal podomere of penduncle longer than combined lengths of distal two podomeres, latter two subequal in length, and distal podomere not quite reaching base of lateral spine on antennal scale; anterolateral spine of first podomere reaching about midlength of second podomere of peduncle; lateral flagellum about five and mesial one about three times longer than carapace. Antenna with peduncle as illustrated, basal segment with ventrolateral spine, flagellum about eight times longer than carapace. Antennal scale slightly less than three times longer than broad with lateral margin almost straight. First pereiopod overreaching antennal scale by length of dactyl, latter subequal in length to palm of chela; carpus slightly more than twice length of chela, and 1.1 times longer than merus. Second pereiopod overreaching antennal scale by length of chela and half that of carpus; chela with fingers slightly longer than smooth palm, former without tubercles or denticles on opposable margins, but both fingers with scattered fine setae and subapical cluster of stiff setae; carpus 1.5 times as long as propodus and about 1.5 times longer than merus, and merus 1.3 times longer than ischium. Third pereiopod overreaching antennal scale by length of dactyl and one fourth that of propodus; propodus slightly more than three times length of dactyl and 1.5 times longer than carpus; latter about one-half as long as merus, and merus 2.6 times longer than ischium. Fourth pereiopod overreaching antennal scale by length of dactyl and three fifths that of propodus; propodus about 4.2 times as long as dactyl and 1.6 times length of carpus; carpus about one-half as long as merus, and latter slightly more than three times length of ischium, Fifth pereiopod overreaching antennal scale by length of dactyl and almost entire length of propodus; propodus, subequal in length to merus, 5.5 times length of dactyl and 1.5 times that of carpus; carpus 3.1 times as long as ischium. First pleopod with exopodite slightly more than twice as long as endopodite. Second pleopod with exopodite 1.2 times as long as endopodite, and latter with appendix masculina reaching distinctly beyond its midlength. Lateral ramus of uropod with straight lateral margin bearing fixed spine and also with slightly longer movable spine at mesial base of latter" (Hobbs et al., 1977).

Habitat: This shrimp lives exclusively in freshwater caves. Distributional records: The reports are only in the Acatlan de Perez Figueroa region in Oaxaca State. This freshwater prawn has been registered in several caves around this area.

Macrobrachium acherontium. Holthuis, 1977.

"The rostrum is high and straight, in the females it is relatively shorter and higher than in the males. It reaches almost to the end of the scaphocerite, sometimes it barely overreaches the end of the antennular peduncle. The upper margin is somewhat convex and bears 8 (rarely 7) to 11 teeth, two or three of which are placed behind the orbit, in the anterior 1/5 to 1/6 of the carapace (rostrum excluded). The dorsal teeth are all very similar and rather evenly spaced. The ventral margin is very convex and usually bears a single tooth in the distal part, sometimes this tooth is altogether absent and rarely there are two teeth. The midrib divides the rostrum in two halves of about equal height, and merges with the orbital margin. The lower orbital angle is lobiform and rounded. The antennal spine stands just below the lower orbital angle at a short distance behind the anterior margin of the carapace. The hepatic spine is as strong as the antennal and stands behind and below it; it is connected with the anterior margin of the carapace by the branchiostegal line. A blunt ridge extends from the antennal spine back to just above the hepatic spine. A hepatic groove is distinct and extends from below the hepatic spine backward. A faint branchiocardiac groove is visible in the poste- rior part of the carapace. The abdomen is smooth. The pleura of the first four somites are broadly rounded, that of the fifth somite is more angularly rounded posteriorly. The sixth somite is about 5/3 as long as the fifth, its pleura are pointed, as is also the posterolateral process; the latter shows a rectangular lobe below the sharp tooth. The telson is 1.2 times as long as the sixth abdominal somite. It bears two dorsal pairs of spines; the anterior pair is placed in the middle of the telson, the posterior is somewhat closer to the anterior pair than to the posterior margin of the telson. This posterior margin ends in a sharp triangular median point and bears two pairs of spines. The spines of the outer pair are short, being about 1/3 of the length of those of the inner pair. Between the inner spines there are several (about 10) strong feathered setae. The eyes have the cornea much reduced, being much shorter and narrower than the eyestalk; it is globular, however, and shows a distinct ocellus. The eye reaches about to the middle of the basal segment of the antennula. The stylocerite is slender and sharply pointed. It reaches to about the middle of the basal segment of the antennular peduncle. The anterolateral angle of the basal segment ends in a sharp tooth, which reaches beyond the middle of the second segment. The anterior margin of the basal segment mediad of the anterolateral spine is convex and is far overreached by the spine. The second and third segments of the peduncle are about of the same length, and much shorter than the first. The outer flagellum has the two rami fused for a short distance, about 7 segments long; the free part of the shorter ramus is much more than 7 times as long as the fused part. The scaphocerite reaches beyond the antennular peduncle and about as far as the tip of the rostrum. It is about three times as long as wide. The outer margin is straight and ends in a sharp tooth, which is far overreached by the lamella, which has the anterior margin broadly rounded, with the inner part sometimes somewhat produced. A lateral spine and a dorsal lobe are present on the antennal peduncle at the base of the scaphocerite. The end of the antennal peduncle does not reach so far as the basal segment of the antennular peduncle. All pereiopods are provided with a pleurobranch, but no other gills, epi- or exopods. The first leg reaches with the fingers or less beyond the scaphocerite. The fingers are slightly longer than the palm. The carpus is some- what less than twice as long as the chela and of about the same length as the merus. The, second pereiopod reaches with the chela and part of the carpus beyond the scaphocerite. It is smooth and slender and no spinules are present. The chela is long and narrow, the fingers are about as long as the palm and slightly narrower. The cutting edges bear a few very small teeth in the extreme proximal part. The carpus is somewhat shorter than the chela, but much longer than the palm, it also is conspicuously longer than the merus. The ischium is slightly shorter than the merus. The third leg reaches with the dactylus and part of the propodus or with only part of the dactylus beyond the scaphocerite. The dactylus is 0.3 to 0.36 times as long as the propodus. The latter is slender and bears a few (about 4) spinules on the posterior margin. The carpus is 1.5 times to twice as long as the dactylus. The merus is distinctly longer and wider than the propodus. The fourth leg reaches slightly farther forward than the third, it surpasses the scaphocerite with half or somewhat less than half the propodus; it is longer and consequently more slender than the third leg. The dactylus is about as long as that of the third leg. The propodus is longer than that of the third leg and 4 to 5 times as long as the dactylus; its posterior margin bears a few more spinules. The carpus is slightly less than 2/3 as long as the propodus, and longer than that of the third leg. The merus is distinctly longer than the propodus and than the merus of the third leg. The fifth leg is the longest of the three walking legs and reaches with half or somewhat less than half the length of the propodus beyond the scaphocerite. The dactylus is only slightly longer than those of the third and fourth legs; it measures 1/5 to 1/4 of the length of the propodus. The latter has several very small spinules on the posterior margin, and in the distal part of that margin some transverse rows of setae. The propodus is almost 1.5 times as long as that of the third leg. The carpus measures about 2/3 of the length of the propodus; the merus is slightly shorter than the propodus. The first pleopod of the male has the endopod ovate with the inner margin concave, it bears no appendices. The endopod of the second pleopod of the male has the appendix masculina about twice as long as the appendix interna, and provided with bristles. The protopod of the uropods has dorsally a sharply acute lobe externally and a blunt lobe medially. The outer margin of the exopod is straight, and ends in an acute tooth. This tooth bear at its inner side a movable spine, which projects beyond the tip of the tooth. The endopod is ovate" (Holthuis, 1977).

Habitat: This freshwater prawn lives exclusively in freshwater caves.
Distributional range: Only known to locality types in Cocona Grootes from Tabasco State.

Macrobrachium sbordonii. Mejia-Ortíz, Baldari & López-Mejia, 2008.

"Medium sized prawn, maximum total length 39.3 mm. Rostrum short, curved anteriorly, tip not reaching the distal border of scaphocerite but surpassing the third article of antennular peduncle; dorsal margin bearing 7 teeth, 1 in postorbital position, two teeth on ventral margin. Live *M. sbordonii* new species is white, without pigment in the body. Carapace smooth, maximum length 17 mm, with antennal spine smaller than hepatic spine. Branchiostegal groove shallow. Abdomen smooth, pleura of first three somites broadly rounded. Posteroventral margin of fourth and fifth pleura acute, all pleura bearing setae on ventral border.

Sixth somite 1.3 times as long as fifth. Telson 1.2 times longer than sixth somite, shorter than uropodal rami; bearing two pairs of dorsal spines, first pair in distal fifth, second pair in middle section of telson; posterior margin broadly triangular bearing two pairs of lateral spines, inner pair 5 times longer than external one, with plumose setae between inner spines. Eyes reduced, cornea with a small apical black point, this point bearing facets. Antennules with acute stylocerite reaching proximal third of first peduncular segment. First antennular segment with acute distolateral spine and concave depression to fit eye. Second antennular segment cylindrical, with sinuous distal margin and lateral row of long setae. Antennae with basicerite bearing large spine on internal margin. Scaphocerite 2.4 times as long as wide, distolateral spine short, widely separated from distal margin of main blade. First pereiopods slender, smooth, with scarcity tufts of setae on both fingers. Tip of fingers surpassing distal margin of scaphocerite; palm slightly compressed, as long as dactylus; carpus 4.5 times palm length, 1.12 times merus length. Second pair of pereiopods slightly equal in size, without spines. Palm cylindrical, 3 times as long as wide, with disperse tufts of setae, 0.75 times dactylus length; carpus 2 times palm length, 1.09 times as long as merus; ischium 0.72 times merus length. Fingers not gaping, elongate, cutting margins covered with tufts of setae, fixed finger and dactylus without teeth. Propodus and dactylus of third pereiopod with several tufts of short setae. One row of three tufts of setae on inner border of propodus, distal pair of tufts of setae on articulation with dactylus and two rows of 7 and 9 movable spines on inner margin of propodus 1 pairs distally in articulation with dactylus. Propodus 2.5 times length of dactylus, 2 times carpus length. Fourth pereiopods, sparsely pilose; propodus 3 times dactylus length, 1.85 times as long as carpus; with one row of three movable spines on inner border of propodus, one pair of spines and tufts of setae on propodus–dactylus articulation. Fifth pair of pereiopods the longest. Propodus and carpus pilose; one longitudinal row of 4 movable spines, 1 spine on propodus–dactylus articulation; propodus 4 times dactylus length, 1.6 times carpus length. Appendix masculina 1.7 times length of appendix interna, inner margin with 12 pairs of spines" (Mejía-Ortíz et al., 2008).

Habitat: Freshwater caves. Distributional records. It is known only from a locality type in Sotano La Lucha, Limits between Oaxaca and Chiapas States.

Troglomexicanus perezfarfantae. Villalobos, Alvarez & Illife, 1999.

"Carapace smooth without antennal hepático branchiostegal spines. Anterolateral margin with rounded suborbital angle. Rostrum short, not reaching distal end of first antennular segment, curving upward. Hepatic and branchial grooves of carapace visible. Abdomen smooth, somites 1-5 with rounded pleura, pleura of sixth somite with posterior portion falcate. Telson shorter than external branch of uropods, longer than internal branch, posterior margin rounded, bearing 3 or 4 spines laterally and 2 pairs on distolateral angles, external pair longer. Antennular peduncle shorter than antennal scale, stylocerite short and blunt with deep dorsolateral groove on first article. External flagellum divided into 2 rami from fourth or fifth basal segment onward; internal ramus short with 10-12 segments. Antenna with antennal flagellum 3 times as long as total length of shrimp. Antennal scaphocerite reaching be- yond antennular peduncle. First pair of pereiopods shorter than second pair. Ischium short, distal portion wider, bear- ing setae on internal margin. Merus longer than carpus. Second pair of pereiopods subequal, slender, longer than total length of shrimp. Surface of chela smooth with scat- tered setae, movable finger two-thirds length of cheliped, cutting edges of fingers smooth, without teeth. Carpus 2.5 times length of palm and longer than merus. Remaining 3 pairs of pereiopods increasing in length posteriorly. Second pair of pleopods with an appendix interna more than half length of endopod. Uropods with external ramus longer than internal one. Exopod bearing 2 spines on laterodistal angle. Endopod as long as telson" (Villalobos et al., 1999).

Habitat: This shrimp lives in freshwater caves. Distributional records: Reported only in Sierra del Abra caves in the northeast of Mexico.

Troglomexicanus tamaulipasensis. Villalobos, Alvarez & Iliffe, 1999.

"Rostrum slender, short, unarmed dorsally or ventrally, proximal half straight, distal half curving downward, ending in sharp tip, reaching slightly beyond ocular peduncles and proximal third of first antennular segment. Carapace smooth, with shallow hepatic depression and without antennal branchiostegal, or hepatic spines. Carapace with anterior margin projected next to rostrum, partially covering ocular peduncles; anterolateral margin rounded at inferior orbital angle; posterior margin formed laterally by broadly rounded expansion overlapping with first abdominal somite. Eyes reduced, without cornea. Abdomen smooth. Anterior margin of pleura of first somite overlapping carapace. Second somite with pleura broadly rounded and with shallow depression along inferior border. Somites 3-5 with posterolateral margin of pleura rounded. Sixth somite slightly longer than second, dorsally projecting on top of anterior portion of telson, with rounded notch on posterolateral margin of pleura at articulation of uropod. Telson shorter than internal pair of uropods; with 2 pairs of small, articulated spines on dorsal surface next to lateral margin. Posterior pair next to posterior margin, anterior pair next to posterior one on distal fifth of length of telson. Posterior margin rounded, bearing 14 long, plumose setae, with 2 spines on each posterolateral angle, external one shortest. Antennular peduncle shorter than antennal scale. Stylocerite strong, with subacute apex reaching distal fourth of first antennular segment. First antennular segment flattened and twice as long as second. Second segment with rounded, deep notch on anterior margin; external flagellum divided into 2 rami from fourth basal segment onward, short ramus with 16 segments. Antenna with scaphocerite reaching beyond antennular peduncle, twice as long as broad, with lateral margin ending distally in small acute spine, not reaching distal margin. First pair of pereiopods slender, shorter than second pair; carpus reaching beyond distal end of scaphocerite, longer in males than in females. Ischium short, less than half of merus length, about half length of carpus and chela; proximal half of ventral border bearing wide notch; tufts of setae on distal half. Merus longer than carpus and chela, dorsal and ventral margins parallel, with tuft of setae ventrally on proximal half. Proximal two-thirds of carpus slender, becoming wider distally, of about same length as chela, with scattered setae dorsally and distal tuft of setae ventrally. Chela slightly thicker than distal portion of carpus; palm three-fourths length of fingers, with ventral tuft of setae on proximal portion. Fingers slender, straight, leaving small gap when closed, with tufts of setae becoming denser distally. Cutting edges of fingers chitinized, devoid of teeth. Tips of fingers sharp, corneous, with tuft of setae. Second pair of pereiopods subequal, elongate, devoid of setae except for 1 tuft on tip of each finger; distal third of merus reaching beyond scaphocerite. Length ratios of segments as follows. Ischium shortest, 0.6-0.8 times length of merus, 0.5-0.7 times length of carpus, and 0.6-0.8 times length of dactyl. Merus shorter than carpus and of same length as dactyl. Carpus of same length as dactyl or slightly longer. Chela longer than carpus, 1.28-1.46 times its length; palm slightly thicker than carpus, about half length of dactyl. Fingers thin, elongate, with sharp corneous tips; cutting edges devoid of teeth, closing completely. Remaining pereiopods thin, increasing in length from third to fifth pair; smooth with sparse short setae near articulations of segments, along posterior margin of propodus, and on middorsal portion of dactyl. In all pairs, dactyl being shortest segment, simple, ending in sharp tip. Third pair of pereiopods with distal margin of carpus reaching beyond scaphocerite. Ischium slightly more than half length of merus and propodus, shorter than carpus, and longer than dactyl. Merus 1.5 times length of carpus, 3 times length of dactyl, and about same length as propodus. Carpus shorter than propodus and twice length of dactyl. Propodus 3 times length of dactyl. Fourth pair of pereiopods with distal third of carpus reaching be- yond scaphocerite. Ischium about half length of merus and propodus, shorter than carpus, and longer than dactyl. Merus 1.5 times length of carpus and 3 times that of dactyl, about same length as propodus. Carpus shorter than propodus and twice length of dactyl. Propodus 3 or more times length of dactyl. Fifth pair of pereiopods with distal third of carpus reaching beyond scaphocerite. Ischium slightly less than half length of merus and propodus, shorter than carpus, and twice length of dactyl. Merus 1.3 times length of carpus, 3 times length of dactyl, and slightly shorter than propodus. Carpus shorter than propodus, more than twice length of dactyl. Propodus 4 times length of dactyl. Pleopods with well-developed rami, except for first pair in both sexes where endopod reduced, without appendix interna. In male, second pair with appendix masculina slender, long, reaching distal third of endopod; bearing setae distally, 4 located on apex; appendix interna, slim, shorter than appendix masculina. Uropods with both rami of subequal length, reaching beyond tip of telson with distal fourth. Protopod with external lobe ending in acute spine and medial lobe rounded. Endopod bearing setae on posterior margin and on distal portion of lateral margins, posterior margin rounded. Exopod with straight external lateral margin ending in unmovable spine, with small, movable, sharp spine next to it on mesial margin. Diaeresis incomplete, appearing as border reaching middle of exopod surface; posterior margin broadly rounded, bearing setae" (Villalobos et al., 1999).

Habitat: This shrimp lives in freshwater caves. Distributional records: Reported only in Sierra del Abra caves in the northeast of Mexico.

Troglomexicanus huastecae. Villalobos, Alvarez & Iliffe, 1999.

"Rostrum short, sharp, slightly oriented upward, reaching beyond distal margin of ocular peduncles and proximal third of first segment of antennular peduncles; dor- sal and ventral margins devoid of teeth. Carapace smooth, without antennal, branchiostegal, or hepatic spines; with shallow branchiostegal depression. Anterior portion of carapace produced next to rostrum, partially covering ocular peduncles; anterolateral margin with inferior orbital angle slightly marked by smooth, rounded projection. Posterolateral margin of carapace ending in broadly rounded section overlapping first abdominal somite. Eyes reduced, ocular peduncles partially exposed, cornea absent. Surface of abdomen smooth. Anterolateral margin of pleura of first somite overlapping posterior margin of carapace. Second somite with pleura broadly rounded, with shallow depression on middle section of inferior margin. Third to fifth somite with posterolateral angle of pleura rounded and projected posteriorly. Sixth somite as long as second somite, with rounded notch on posterolateral margin of pleura at articulation of uropod, angles limiting rounded notch subacute; dorsal posterior margin slightly projected on top of anterior portion of telson. Telson shorter than internal rami of uropods, 2 small, movable pairs of sharp spines dorsally along lateral margin; first pair at approximately four-fifths length of telson, second pair smaller, next to posterior margin. Posterior margin of telson almost straight, with 14 long, plumose setae, and with pair of spines on each posterolateral angle, external one smaller. Antennular peduncle shorter than antennal scale; stylocerite strong, apex subacute, reaching distal fourth of first peduncular segment. Second segment with deep, rounded notch on anterior margin. External flagellum divided into 2 branches from fourth basal segment onward, short branch with 16 segments. Antennae with scaphocerite reaching beyond antennular peduncles with distal fourth, more than twice as wide as long, lateral margin ending in small sharp spine, not reach- ing distal margin. First pair of pereiopods slender and shorter than second pair. Ischium short, slightly more than half length of merus, carpus, and chela; ventral border with wide proximal notch and tufts of setae along distal half. Merus slightly longer than carpus and chela, dorsal and ventral margins parallel, tuft of setae on proxi- moventral margin. Proximal two-thirds of carpus slender, becoming wider distally, as long as chela, scattered setae on dorsal margin, and tuft of setae on distoventral margin. Chela thicker than distal portion of carpus, palm shorter, about half length of fingers, with proximoventral tuft of setae. Fingers slender, straight, with scattered setae becoming denser distally; leaving narrow space when closed, cutting borders smooth without teeth, tips corneous, with tuft of setae. Second pair of pereiopods subequal, long, slender, with scattered setae; reaching beyond distal margin of scaphocerite with distal third of merus. Length ratios of segments as fol- lows. Ischium shortest, 0.76 times length of merus, 0.63 times length of carpus, and 0.72 times length of dactyl. Merus shorter than carpus (0.8) and slightly shorter than dactyl (0.95). Carpus longer than dactyl (1.14). Chela 1.4 times length of carpus; palm slightly thicker than carpus, more than half length of dactyl. Fingers thin, elongate, with sharp, cor- neous tips; cutting edges devoid of teeth, closing completely. Remaining pereiopods slender, increasing in length from third to fifth pair; smooth with sparse short setae near articulations of segments, along posterior margin of propodus, and on mid-dorsal portion of dactyl. In all pairs, dactyl shortest segment, simple, ending in sharp tip. Third pair of pereiopods with ischium more than half length of merus and propodus, 0.9 times length of carpus, 1.75 times length of dactyl. Merus 1.5 times length of carpus, 3 times as long as dactyl, and of about same size as propodus. Carpus shorter than propodus and twice length of dactyl. Propodus 3 times length of dactyl. Fourth pair of pereiopods with ischium more than half length of merus, shorter than carpus, half length of propodus, and 1.6 times length of dactyl. Merus 1.5 times length of carpus, 2.7 times length of dactyl, and slightly shorter than propodus. Carpus shorter than propodus and about twice length of dactyl. Propodus 3.2 times length of dactyl. Fifth pair of pereiopods with ischium more than half length of merus, shorter than carpus and propodus, and 1.6 times length of dactyl. Merus 1.3 times length of carpus and 2.8 times length of dactyl, shorter than propodus. Carpus more than half length of propodus and more than twice length of dactyl. Propodus 3.8 times length of dactyl. Pleopods with well-developed rami, except for first pair where endopod reduced without appendix interna. Uropods with internal branch slightly shorter than external one, reaching beyond. Distal margin of telson with distal fourth. Protopod with external lobe forming sharp spine and internal lobe rounded. Endopod with setae along posterior margin and distal portion of lateral margins, posterior margin rounded. Exopod with lateral margin straight, ending distally in unmovable spine, with internally small, sharp, movable spine; diaeresis incomplete, appearing as border reaching middle of dorsal surface of exopod; posterior margin broadly rounded, bearing setae" (Villalobos et al., 1999).

Habitat: This shrimp lives in freshwater caves. Distributional records: Reported only in Sierra del Abra caves in the northeast of Mexico.

Cryphiops sbordonii. Baldari, Mejía-Ortíz & López-Mejia, 2010.

"Medium sized prawn, maximum total length 54.5 mm. Rostrum short, straight, tip not reaching the distal border of scaphocerite but almost reaching the the third article of antennular peduncle; dorsal margin bearing 8 teeth, lack teeth in postorbital position and on ventral margin. Live *Cryphiops sbordonii* new species is white, without pigment in the body. Carapace smooth, maximum length 25 mm, with only antennal spine. Branchiostegal groove shallow. Abdomen smooth, pleura of first three somites broadly rounded. Posteroventral margin of fourth and fifth pleura acute, all pleura bearing setae on ventral border. Sixth somite 1.5 times as long as fifth. Telson 1.4 times longer than sixth somite, shorter than uropodal rami; bearing two pairs of dorsal spines, first pair in distal fifth, second pair in middle section and between these of telson have a single spine on left side; posterior margin broadly triangular bearing two pairs of lateral spines, inner pair 5 times longer than external one, with plumose setae between inner spines. Eyes reduced, cornea with a small apical black point, this point bearing facets. Antennules with short stylocerite on the proximal third of first peduncular segment. First antennular segment with acute distolateral spine and concave depression to fit eye. Second antennular segment semi-cylindrical, with sinuous distal margin and lateral row of long setae. Antennae with basicerite bearing short spine on internal margin. Scaphocerite 2.4 times as long as wide, distolateral spine short, widely separated from distal margin of main blade. First pereiopods slender, smooth, with scarcity tufts of setae on both fingers. Palm surpassing distal margin of scaphocerite; palm slightly compressed, as long as dactylus; carpus 1.75 times palm length, 1.12 times merus length. Second pair of pereiopods slightly equal in size, without spines. Palm semi- cylindrical, 3.3 times as long as wide, with disperse tufts of setae, 0.8 times dactylus length; carpus 1.19 times palm length, 0.8 times as long as merus; ischium 0.9 times merus length. Fingers not gaping, elongate, cutting margins covered with tufts of setae, fixed finger and dactylus without teeth. Propodus and dactylus of third pereiopod with several short setaes. One row of seven spines on inner border of propodus. Propodus 3 times length of dactylus, 2.05 times carpus length. Fourth pereiopods, sparsely pilose; propodus 3.4 times dactylus length, 1.87 times as long as carpus; with one row of nine movable spines on inner border of propodus, one pair of setae on propodus–dactylus articulation. Fifth pair of pereiopods the longest. Propodus and carpus pilose; one longitudinal row of 12 movable spines the last four closed to distal section, 1 spine on propodus–dactylus articulation; propodus 4 times dactylus length, 2.1 times carpus length. Appendix masculina 2 times length of appendix interna, inner margin with 10 pairs of spines" (Baldari et al., 2010).

Habitat: This shrimp lives in freshwater caves. Distributional records: Only known to locality types in the Las Margaritas region from Chiapas State.

Cryphiops luscus. Holthuis, 1974.

"Eyes with reduced but pigmented corneal area. Rostrum with dorsal and ventral teeth. Carapace with antennal spine but lacking supraorbital, hepatic, and branchiostegal spines. Telson with 2 pairs of dorsal and 2 pairs of posterior spines, and posterior margin with numerous setae between mesial pair of spines. Mandible with 3-segmented palp. Second maxilliped with podobranch, third with pleurobranch. Second pereiopods very robust and spinulate; third to fifth with simple dactyl. Propodus of fifth pereiopod with fringe of hair on posterodistal margin. Appendix interna absent on first pleopod; appendix masculina distinctly overreaching appendix interna (Hobbs & Hobbs, 1976).

Habitat: This species lives in freshwater caves. Distributional range: Known only in the locality type in the Trinitaria Region, Chiapas, Mexico.

Cambaridae Hobbs, 1942. *Procambarus xilitlae.* Hobbs & Grubbs, 1982.

"Albinistic, eyes without pigment or faceted cornea. Rostrum with small marginal spines or tubercles; median carina absent. Carapace with cervical tubercle cephaloventral to rwo or minute tubercles flanking caudal margin of cervical groove. Areola 4.7 to 5 times as long as broad, constituting 36.5 to 37.6 percent of total length of carapace (46.1 to 46.6 percent of postorbital carapace length). Suborbital angle absent. Postorbital ridge with cephalic spine, lacking posterodorsal ones. Hepatic area with very few small tubercles. Antennal scale about 1.8 times as long as wide, broadest slightly distal to midlength, Ischia of third and fourth pereiopods of second form male with small hooks, neither approaching corresponding basio-ischial articulation not opposed by tubercle on basis; caudomesial angle of coxa of fourth pereiopod with prominent boss triangular in mesial aspect, and coxa of fifth with much smaller one compressed in longitudinal plane of body. First pleopods of second form male reaching coxae of third pereiopods, slightly asymetrical, provided with subapical setae, and bases of paired members widely separated; distal extremity bearing prominent conical mesial process, largest of terminal elements, directed caudolaterally and somewhat distally; cephalic process represented by subangular hood flanking cephalic surface of centrally located central projection, and caudal process consisting of swollen ridge on caudolateral extremity of shaft of appendage. Annulus ventralis hinged to sternum inmediately anterior to it, not freely movable, suboval in outline, almost twice as wide as long, rather strongly arched ventrally, and bearing sigmoid sinus extending along most of median line of annulus, tongue and fossa not clearly defined. Sternum anterior to annulus weakly tuberculate or with causally directed prominences flanking median line. Postannular sclerite triangular, its base about twice as broad as long and abouth 0.6 as wide and 0.5 as long as *annulus*" (Hobbs & Grubbs, 1982).

Habitat: Species exclusively from freshwater caves. Distributional records: Known from locality types in the Xilitla Region, San Luis Potosí.
Procambarus cavernicola. Mejía-Ortíz, Hartnoll & Viccon-Pale, 2003.

"Body unpigmented; eye reduced with setae on the tip, facets undefined, cornea with small area of purple pigment. Rostrum excavate, with narrow margins moderately converging to base of acumen, this delimited each side by spine, reaching middle portion of third antennular podomere, 26.2 to 38.4% ($\bar{x}$ 32.5% n= 22) of rostrum length. Areola 3.1 to 6.7 ($\bar{x}$ 4.4 n= 22) times as long as wide, 30.4 to 36.1% ($\bar{x}$ 32.9% n= 22) of carapace length, 43.3% to 49.6% ($\bar{x}$ 46.4% n= 22) of postorbital carapace length, with 5 to 8 punctuations across the narrowest part. Single cervical and branchiostegal spine on each side of carapace. Postorbital ridge strong, groove weak, cephalic margin with small spine or acute tubercle. Antennal scale 1.5 to 2.4 ($\bar{x}$ 2.0) times longer than wide, greatest width at midlength; lateral margin thickened, terminating distally in spine. Antennal flagellum with tip extending beyond caudal margin of telson when flagellum addressed. Chelipeds as long as body, mesial surface of palm of chela with 23 tubercles in irregular row; surface of chela covered with small regularly distributed tufts of short setae, fingers as long as palm; both fingers with longitudinal ridges on ventral and dorsal surfaces. Ischium of third pereiopod armed with acute hook, extending beyond articulation with basis. Epistome semitriangular, covered with setae, lateral angles rounded. First pleopods of Form I male of equal length, tips reaching coxae of third pereiopod, with 2 rows of subapical setae. Mesial process largest of terminal elements, long, acute, directed distally and inclined slightly laterally; central projection triangular directed cephalically; cephalic process reduced, cephalic shoulder forming convex border. *Annulus ventralis* subovate, cephalic half with median trough, flanked each side by rounded ridge; sinus deep, originating at caudal end of trough, curving sharply dextrally and making hairpin turn before curving to dissect caudal margin slightly dextral to midline. Preannular plate with longitudinal median trough, lateral areas swollen, covered with short setae, postannular plate nearly same size as annulus, oval-shaped, apical surface concave, with blunt tubercles, not in contact with *annulus*" (Mejía-Ortíz et al., 2003).

Habitat: Known as being from freshwater caves. Distributional range: Known only from the locality type, the Gabriel Cave in the Acatlán de Pérez Figueroa Region from Oaxaca State.
Procambarus rodriguezi. Hobbs, 1943.

"Body without pigment or pale orange; eyes with small pigment spot but no facets. Areola constituting 35.8 to about 37.5 percent of total length of carapace (43.5 to 44.3 percent of postorbital carapace length). Cervical spine present. Antennal scale broadest distal to midlength. Coxa of fifth pereiopod lacking prominent, sclerotized caudomesial boss. First pleopod of first form male with mesial process strongly flattened, tapering distally, and directed more distally than laterally; central projection lacking accessory tooth; and caudal element lacking cushion-like prominence" (Hobbs, et al., 1977).

Habitat: This crayfish living in subterranean freshwater springs. Distributional range: This crayfish species is known only in the locality type near Cordoba Veracruz in the karstic hills.
Procambarus oaxacae. Hobbs, 1973.

"Body with markedly reduced pigmentation, virtually albinistic but often with slight tan suffusion on abdomen; eyes reduced in size with or without traces of facets, frequently with few ommatidia bearing reddish purple pigment, but with pigmented area not sharply margined. Areola constituting 35.8 to 37.8 percent of total length of carapace (43.7 to 46.1 percent of postorbital carapace length). Cervical spine obsolete. Antennal scale broadest distal to midlength. Coxa of fifth pereiopod lacking corneous caudomesial boss. First pleopod of first form male with mesial process not strongly flattened, tapering distally, and directed more distally than laterally; central projection lacking accessory tooth; and caudal element lacking cushion-like prominence" (Hobbs et al., 1977).

Habitat: This species lives in freshwater caves. Distributional records: Known only in the locality type from San Juan Bautista Valle Nacional in Oaxaca State.

Procambarus reddelli. Hobbs, 1973.

"Body without pigment or with tan suffusion on abdomen; eyes with distinct black pigmented faceted area. Areola constituting 33.2 to 35.7 percent of total length of carapace (42.0 to 44.4 percent of postorbital carapace length). Cervical spine present. Antennal scale broadest distal to midlength. Coxa of fifth pereiopod lacking prom- inent sclerotized caudomesial boss. First pleopod of first form male with mesial process not strongly flattened, tapering distally, and directed more distally than laterally; central projection lacking accessory tooth; and caudal element lacking cushion-like prominence" (Hobbs et al., 1977).

Habitat: Known only as being from freshwater caves. Distributional records: This species is only know in the locality type: A cave near to San Juan Bautista Valle Nacional, from Oaxaca State.

Pseudothelphusidae. Rathbun, 1893. *Villalobosus lopezformenti.* Alvarez & Villalobos, 1991.

"Gonopod straight and strong. In cephalic view, proximal % of gonopod twice as thich as distal. Mesial process semicircular in mesial view, with proximal field of spines; subtriangular in caudal view. In lateral view, gonopod uniformly broad with mesial process protruding distally. Lateral surface with 3 strong spines coming out from lateral crest. Marginal process simple, straight, reaching apex. In an apical view, mesial process very prominent, slightly curved laterally, apex cavity elongated, field of setae in cephalic portion, field of small spines in caudal portion" (Alvarez & Villalobos, 1991).

Habitat: Known only as being from freshwater caves. Distributional records: This species is very abundant in the locality type: San Antonio Cave in the Acatlán de Pérez Figueroa Region, from Oaxaca State.

Typhlopseudothelphusa hyba. Rodriguez & Hobbs, 1989.

"Carapace narrow (cb/cl = 1.46), strongly convex in anterior part. Cervical grooves forming paired, shallow depressions in holotype but absent in topo- typic female. Postfrontal lobes and median groove obsolete, but carapace with slight depression immediately posterior to front; latter, lacking upper margin, rounded and projecting forward, its lower margin bilobed in dorsal view. Lateral orbital angle forming well developed triangular tooth, followed by shallow notch. Margins of orbits and front with row of small granules. Lateral margin of carapace with larger, irregularly placed granules. Eyes reduced, lacking distinct faceted cornea and pigment. Third maxilliped with noticeable impression on external margin of merus, near insertion of palp; exognath 0.61 times length of ischium. All pereiopods extremely slender. Chelipeds subequal in size and shape; palm cylindrical; fingers also very slender, approximately twice as long as palm, and armed with minute teeth; carpus with small, hooked spine on internal margin; merus very elongate, overreaching carapace by half its length, and bearing rows of granules on its three ridges. Third pair of walking legs 1.6 times width of carapace; merus 5.87 times as long as broad, and relations of its podomeres as follows: merus 1, carpus 0.35, propodus 0.66, dactyl 0.74; dactylus with six or seven spines in five rows. Gonopod wide in lateromesial plane, narrow in cephalocaudal plane; distal part strongly bent; gonopore and field of spines directed cephalad; distal margin rounded and armed with prominent conical spines; marginal process cup shaped and directed cephalad; and strong, triangular mesial process disposed in same plane as field of spines" (Rodríguez & Hobbs, 1989).

Habitat: Known only as being from freshwater caves. Distributional record: Only known from San Cristobal de las Casas Region, inhabiting two caves.

Typhlopseudothelphusa mocinoi. Rioja, 1953.

"First gonopod the male straigh and strong; margin straight. Surface bearing apical spines directed cephalically or disposed at right angle to axis of gonopod. Strong mesial process or apical or subapical lobule always present. Lateral suture and margin forming apical lobule, which in some species prolonged beyond apex of gonopod. Eyes reduced, devoid of pigment and lacking facets. Pereiopods very long and slender" (Hobbs et al., 1977).

Habitat: Freshwater caves. Distributional record: Know only in three caves in Chiapas State.

Potamocarcinus leptomelus. Rodriguez & Hobbs, 1989.

"Cervical groove narrow, almost straight, very shallow, reaching margin of carapace. Postfrontal lobes obsolete, their position marked by slight depressions; median groove absent except for notch on upper margin of front. Surface of carapace between postfrontal lobes and front sloping gently ventromesially. Front low, of equal height throughout, bilobed in dorsal view, upper margin defined by series of irregularly placed papillae; lower margin V-shaped in median part, straight in lateral parts. Dorsal margin of orbits sinuous. Lateral orbital angle forming papillated tooth, followed by deep notch; rest of anterolateral margin covered by small, irregularly placed, conical papillae. Surface of carapace covered by very small papillae, invisible to naked eye, particularly in branchial region. Eyes small, with distinct faceted cornea and pigment, of usual shape, not filling or bit. Third maxilliped with conspicuous notch on external margin of merus, near insertion of palp; external margin of ischium convex; exognath 0.59 times length of ischium. As name suggests, all pereiopods conspicuously slender. Chelae only slightly dissimilar in size and shape; palm of larger one very elongate with straight upper and lower margins; fingers not gaping; carpus with strong internal spine preceded by four spinules; merus elongate with row of spinules on internodorsal and interno ventral ridges and smaller spinules on outer ridge. Walking legs very long, length of third pair 1.6 times breadth of carapace, and merus 6.3 times as long as broad; relations between podomeres as follows: merus 1, carpus 0.33, propodus 0.54, dactylus 0.54. Dactylus with seven spines in upper row and four in two lower ones. Gonopod subcylindrical, straight; apical part bearing triangular marginal process overreaching apex, spiniform mesial process directed distomesially, and bifid cephalic process with apices pointing mesially. Field of spines very narrow, with few setae placed on distal surface of ápex" (Rodríguez & Hobbs, 1989).

Habitat: Only live in freshwater caves. Distributional records: Only known from locality types in the Zongolica region, from Veracruz State.

Odontothelpusa monodontis. Rodríguez & Hobbs, 1989.

"Cervical groove very shallow and wide, not reaching margin of carapace. Postfrontal lobes wide, flat, clearly delimited anteriorly, surface covered by flat papillae not visible to naked eye; median groove well marked between lobes, but not anterior to them. Surface of carapace between frontal lobes and front flat. Upper border of front well marked by overhanging papillae and conspicuous notch in middle; lower margin slightly sinuous and lying slightly anterior to upper one; front high and excavate. Anterior margin of carapace with notch behind orbit, about 5 small papillae situated between notch and cervical groove, and approximately 15 papilliform teeth on rest of lateral margin. Third maxilliped conspicuously wide; merus 1.43 times as wide as long; exognath 0.93 length of ischium of endognath. Gonopod slender with apical part bent laterally at angle of 45 degrees; apex con- sisting of flat quadrangular lobe with long spur on proximomesial angle and papilliform tubercle and small finger-like tubercle on cephalic surface; field of spines narrow, slitlike" (Rodríguez & Hobbs, 1989).

Habitat: Freshwater caves. Distributional records: Known only from locality type: Cocona Grootes, in Tabasco, Mexico.

Pseudothelphusa sonorae. Rodríguez & Smalley, 1969.

"The forehead does not have a superior margin, but forms a smooth curve. The forehead is not vertical and therefore is visible in dorsal view. The lower margin of the forehead is sinuous, with a deep indentation in between. The medium groove is shallow. The epigastric lobes are low, and would not be visible except for the shallow depressions located in front of them. The anterolateral teeth are small and blunt. There is a deep indentation of smooth edge immediately posterior to the external orbital angle. The chellipeds are unequal, the left one is larger in both type specimens. The palm is bulging, with the lower margin convex only at the base and the distal 3/4 forming a concave and shallow arch. The teeth of the chelicerae are alternately large and small, sometimes with small teeth among the large teeth. The internal margin of the carpus has a single large internal tooth, with four low and blunt teeth, and a tubercle on the inner margin. The inner edge of the grouper is covered with low teeth arranged in a very irregular sequence. The gonopod resembles that of other northern Mexican species of the subgenus Pseudothelphusa. There is not a notch on the mesial lobe at the tip of the gonopod, but it carries a crest with small teeth as in *P. lophophallus*. The mesial lobe does not curve sharply on the gonopod. The apex does not appear as high as in *P. jouyi*. The lateral lobe is short, sharp and without deep notches or projections" (Rodríguez & Smalley, 1969).

Habitat: Freshwater cave. Distributional record: Only known from old mines in Alamos Sonora.

Trichodactylidae. Milne Edwards, 1853. *Rodriguezia mensabak*. Cottarelli & Argano, 1977.

"Carapace and pereiopods completely depigmented. Carapace suborbicular, slightly convex; upper surface smooth with pores; internal organs visible by transparency; gastric and hepatic regions prominent postfrontal lobes slightly prominent; antero-lateral margin devoid of teeth. Front slightly bilobed, depressed and inclined downward; postero-lateral and inferior margins of front bordered but small tubercles. Eyes devoided of cornea, eyestalk short and stout, little mobile. All abdominal segments distinct in male and female; male abdomen triangular, outer margin straight last segment widely rounded, approximately 0.56 as long as broad. Third maxilliped with distal margin of merus sinuous. Chelipeds subequal; fingers with inconspicuous tubercles. Legs extremately long and slender; dactylus covered with minute setae. First gonopod straight, wider at basal portion, constricted at 2/3 of its length and slightly expanded in distal third; strong conical spines over distal portion, patch of spinules over basal mesial angle and few setae over lateral border; gonopore long, suboval. Second gonopods considerably shorter than first (approximately ½). Terminal article considerably shorter (approximately ½) than first, slightly recurverd, subtriangular and acuminate, with few denticles" (Rodríguez, 1992).

Habitat: Freshwater caves. Distributional records: Only known to Chiapas caves.

Rodriguezia villalobosi. Rodríguez & Manrique, 1967.

"Carapace suborbicular; upper surface with pores and small granules, not visible to naked eye, rest smooth and polished; gastric region more prominent than rest, hepatic slightly elevated; frontal region concave, progressively sloping downwards to margin of front; postfrontal lobes inconspicuous; cardiac, branchio-urograstric and urogastric grooves wide and shallow; oval metagastric region defined by these grooves. Lateral margins angled, devoided of teeth; postero-lateral ridge of carapace bent mesially in posterior end, continuing in thinner ridge parallel to posterior margin. Front moderately bilobed. All abdominal segments distinct in male and female; male abdomen subtriangular, wide, outer margin slightly concave, last segment triangular, pointed. Chelipeds strongly unequeal, largest with palm swolled, inferior margin sinuous; fingers gaping, with rows of darks points on external surface; merus with strong distal spine on the upper border. Legs not slender, dactylus covered by felt-like pubescence, with longer hairs over internal margin; propodus with similar pubescence over distal external portion. First gonopods straight, wider in basal portion, constricted at middle, slightly expanded distal half, with strong conical spines over distal portion; gonopore long, suboval. Second gonopods considerably shorter than first (approximately ½), terminal article considerably shorter (approximately ½) than first, proximally wide, long acuminate distally" (Rodríguez, 1992).

Habitat: Freshwater caves. Distributional records: Only known to Chiapas caves.

Rodriguezia adani. Alvarez and Villalobos, 2018.

"Dorsal surface of carapace smooth; in dorsal view suborbicular, slightly wider than long; branchial, gastric, cardiac regions faintly marked; lateral margin complete, devoid of teeth; posterior margin slightly arched, almost straight; posterolateral and anterolateral angles rounded; anterior margin bilobed, central portion slightly depressed, laterally with small rounded projections that limit orbits. In frontal view, superior border of carapace formed by minute tubercles; median portion elevated; laterally, portion between insertion of antennule and antenna slightly depressed, portion around orbits elevated. Internal orbital angle weakly marked, occlusive tooth rounded, buccal angle oval-shaped. Abdomen of male and female with all segments distinct. Male abdomen triangular, last segment with posterior margin rounded, twice as wide as long. Female abdomen broadly rounded, partially covering thoracic sternites, last segment 4 times as wide as long. Antennule and antenna, typical of genus. Eyes absent, ocular peduncle strongly reduced, not movable, appearing as blunt plug occupying most of orbit. Third maxilliped approximately rectangular, ischium slightly wider than merus; exopod shorter than ischium and merus combined, flagellum with eight articles. Chelipeds subequal in size, carpi with tooth on internal dorsal margin, chelae with dorsal tooth on proximal third, fingers not gaping. Major chela with swollen palm, twice as long as high; fingers about one third length of chela, cutting edges with low irregular teeth, tips acute; in lateral view, movable finger approximately triangular. Minor chela 2.4 times as long as high, fingers about half length of chela. Pereiopods 2–5, elongate, representing 1.84, 2.26, 2.48, 2.21 times carapace breadth. Propodi and dactyli with fine pubescence, more dense on the latter. First gonopod straight, slender, proximal third wider. Middle constriction, typical of genus, faintly visible in lateral and mesial views. Field of spines covering distal third except for lateral surface; all spines similar in size and shape. Apical pore suboval, drop-shaped. Second pleopod less than half length of first, distal portion slender, acute" (Alvarez & Villalobos 2018).

Habitat: Exclusively in freshwater caves
Distributional records: Only known from locality type Grutas de Agua Blanca, Macuspana Tabasco.
Avotrichodactylus bidens. Bott, 1969.

"Carapace suborbicular; upper surface moderately irregular; in frontal view regularly arched; frontal and orbital regions excavated; lobes distinct, not prominent; branchio-urogastric and anterior portion of branchio-cardiac grooves deep, well marked, urogastric and posterior part of branchio-cardiac groove indicated by flat depressions. Postgastric pits absent. Anterio-lateral margin with 2 small, spiniform teeth of approximately equal size behing external orbital angle; interdental space approximately twice space between outer orbital angle and 1st tooth;

posterior-lateral ridge of carapace, not bent mesially in posterior ends as in another species, but parallel to lower ridge located over 5th coxa, ends away from postero-lateral angles of carapace. Margin of front slightly concave. Orbits suborbicular; orbital suture absent or represented by small depression; lower orbital margin papillated; inner orbital angle blunt; occlusive orbital tooth rounded, small, located close to inner orbital angle, but not continuous with it; outer orbital not prominent, its border in contact with orbital margin; margin between outer orbital angle and first tooth of carapace forms rounded lobe; buccal angle smooth. Front advanced. Hiding epistome in dorsal view; anterior surface of front sunk, low in middle, thin on sides; margin over each antennular fossa more or less straight; antennular septum sunken; epistome moderately high; inclined forwars. Third to figth abdominal segments fused in both male and female. Basal article of antenna with outer lobe prominent and narrow; ischium of third maxilliped with deep groove. Chelipeds strongly unequal; larger chela of male with upper border of hand strongly arhed and lower sinuous, fingers with elongated gap between them, teeth small, regularly placed along cutting edges; carpus with inner margin produced in sharp hooked spine; merus with apical spine on upper border, another at distal angle of latero-interior margin and another at inner margin. Legs with long coarse hairs on lower margin of dactylus; similar hairs cover all the lower margin of propodus in 5th, ¾ in 2nd, ½ in third and only distal angle in 4th pereiopod; upper margin of propodus and dactylus with shorter sparce hairs, rest of surface covered by felt-like pubescence; claws of dactyli very short, with indistinct longitudinal carinae, 1 upper 2 lateral and 2 inferior" (Rodríguez, 1992).

Habitat: Freshwater caves. Distributional records: Only known to locality type: Cueva de Villa Luz in Tabasco State.

EVOLUTIONARY REMARKS

It is evident that the diversity of species and families of crustaceans is high due to the complexity of the environment and the ability to provide them protection. Likewise, the forms and adaptations and strategies to survive in these environments have allowed them to be successful in their development (Figure 1). While we have an important group living in the anchihaline environments of the Yucatan Peninsula with a recent geological origin, there are other species, totally freshwater and inhabiting spaces with much more development time. This, coupled with past transgressions and marine regressions have allowed the existence of predominantly marine groups in caves that are totally freshwater today.

Mexico currently has 40 species of cave decapods registered so far in freshwater environments and anchihaline environments, with the Palaemonidae Family being best represented with 10 species, while the Procarididae, Anchialocarididae and Barbouriidae families are those that present one species (Table I). Of course, one of the characteristics of the cave species are their high endemism and in the crustaceans this is not the exception because there are species that are only known in their type locality. It turns them into a great ecological value because apart from being so specialized, they present a microdistribution in their own region together with another species such fishes, echinoderms, worms (Mejía-Ortíz et al., 2007). The adaptations that these animals present are the reduction of the eyes, the enlargement of the ambulatory and sensory structures, the increase in their sensory capacity, the decreased oxygen consumption rate, the increase in their parental care; its nutritional efficiency, the development of specialized structures for the storage of lipids, the setae development in the apical areas of the eyes, and the relationship with sulforeductive chemosynthetic bacteria for survival (Mejía-Ortíz, 2005; 2008). Likewise, the evolutionary forces that shape these adaptations are identified as food energy supply, a biannual or annual hydrological regime, and the time of colonization. Obviously, the organisms listed here in this work show a series of adaptations to cave life linked to the availability of energy as the greatest evolutionary force in caverns (Mejía-Ortíz 2015).

Also, it is important to mention that after many years of exploration and advances in the knowledge of cave decapods, there are still areas that researchers have not reached and organisms continue to be found that have not yet been classified. It is important to mention that efforts must continue to first understand the subterranean biodiversity, but also to understand the functioning of these very complex and so little accessible environments.

ACKNOWLEDGMENTS

This work is the outcome of project Biodiversity in the caves of Quintana Roo; and the author gives thanks to CONACYT for supporting the exploration in Mexico around the karstic areas by funding the project Sistemática molecular de los langostinos del género *Macrobrachium* con desarrollo larval abreviado en el sureste de México y su relación con Guatemala y Belice supported by CONACYT-258494 (Fondo Sectorial de Investigación para la Educación). Also the author gives thanks to all members of the Speleological Circle of Mayab by their contributions in the explored caves.

REFERENCES

Alvarez, F. and Villalobos, J. L. (1991). A new genus and two new species of freshwater crabs from Mexico, *Odontothelphusa toninae* and *Stygothelphusa lopezformenti* (Crustacea: Brachyura: Pseudothelphusidae. *Proceeding of the Biological Society of Washington*, 104(2): 288-294.

Alvarez, F. and Villalobos, J. L. (1998). Six new species of freshwater crabs (Brachyura: Pseudothelphusidae) from Chiapas, Mexico. *Journal of Crustacean Biology*, 18(1): 187-198.

Alvarez, F., Iliffe T. M. and Villalobos, J. L. (2005). New species of the genus *Typhlatya* (Decapoda: Atyidae) from anchialine caves in Mexico, The Bahamas, and Honduras. *Journal of Crustacean Biology*. 25(1): 81-94.

Alvarez, F. and Illife, T. M. (2008). Anchialine faune from Yucatan Peninsula. In: Rodríguez-Almaráz G. y F. Alvarez (Eds). *Crustaceans from Mexico; Currently Known.* Universidad de Nuevo León-Instituto de Biología UNAM.

Alvarez, F., Iliffe, T. M., González B. and Villalobos, J. L. (2012). *Triacanthoneus akumalensis*, a new species of alpheid shrimp (Crustacea: Caridea: Alpheidae) from an anchialine cave in Quintana Roo, México. *Zootaxa*, 3154: 61-68.

Alvarez, F., Illife, T. M., Benitez, S., Brankovits, D. and Villalobos, J. L. (2015). New records of anchialine fauna from the Yucatan Peninsula. *CheckList the Journal of Biodiversity data*, 11(1): 1505.

Baldari, F., Mejía-Ortíz, L. M. and López-Mejía, M. (2010). *Chryphiops sbordonii* (Decapoda: Palaemonidae), a new stygobitic species of freshwater shrimp from southeast of Mexico. *Zootaxa*, 2427: 47-54.

Camacho A. I. (1992). A classification of the aquatic and terrrestrial subterranean environment and their associated fauna. Pp. 57-106. In: Camacho A. I. (Ed). 1992 *The natural history of Biospeleology*. Museo Nacional de Ciencias Naturales. CSIC, Madrid.

Chace, F. A., Jr. (1972). The shrimps of the Smithsonian Bredin Caribbean Expeditions with a sumary of the West Indian shallow wáter species (Crustacea: Decapoda: Natantia). *Smithonian Contribution to Zoology*, 98: 1-179.

Escobar-Briones, E., Camacho M. E. and Alcocer, J. (1997). *Calliasmata nohochi*, new species (Decapoda: Caridea: Hyppolytidae), from anchialine cave systems in continental Quintana Roo, Mexico. *Journal of Crustacean Biology*, 17(4): 733-744.

Espinasa-Pereña, R. 1994. *Origin and distribution of caves in Mexico*. Sciences, Faculty of Sciences, UNAM; 36: 44-49.

Giraldes, B. W., Coelho Filho, P. A., Coelho P. A. and Anker, A. (2012). Confirmation of the presence of *Janicea antiguensis* (Chace, 1972) (Decapoda: Barbouridae) in northeastern and eastern Brazil. *Nauplius*, 20(2): 171-178.

Hart, C. W. Jr. & Manning, R. B. (1981). The cavernicolous carideans shrimp of Bermuda (Alpheidae, Hippolytidae, and Atyidae). *Journal of Crustacean Biology*, 1(3): 441-456.

Hobbs, H. H. III and Hobbs, H. H. Jr. (1976). On the troglobitic shrimps of the Yucatan Peninsula, Mexico (Decapoda: Atyidae and Palaemonidae. *Smithsonian Contribution to Zoology*, 240: 1-23.

Hobbs, H. H. Jr., Hobbs, H. H. III and Daniel, M. A. (1977). A review of the troglobitic decapod crustaceans of the Americas. *Smithsonian Contributions to Zoology*, 244: 1-183.

Hobbs, H. H. Jr. and Grubbs, A. G. (1982). Description of a new troglobitic crayfish from Mexico and a list of Mexican crayfishes repored since the publication of the Villalobos Monograph (1955) (Decapoda, Cambaridae). *Association Mexican Cave Studies Bulletin*, 8: 45-50.

Holthuis, L. B., (1977). *Cave shrimps (Crustacea Decapoda, Natantia) from Mexico*. Accademia Nazionale dei Lincei, Quaderno 171: 173-195.

Holthuis, L. B., (1974). *Bithynops luscus, a New Genus and Species of Cavernicolous Shrimp from Mexico (Crustacea Decapoda, Palaemonidae)*. Quademi Accademia Nazionale dei Lincei, 171:135-142, 2 figures.

Kensley, B., (1988). New species and record of cave shrimps from the Yucatan Peninsula (Decapoda: Agostocarididae and Hippolytidae). *Journal of Crustacean Biology* 8(4): 688-699.

Mejía-Ortíz, L. M., Hartnoll R. G. and Viccon-Pale. J. A. (2003). A new species of stygobitic crayfish from Mexico, *Procambarus cavernicola* (Decapoda: Cambaridae), with a review of cave dwelling crayfishes in Mexico. *Journal Crustacean Biology*, 23(2): 391-401.

Mejía-Ortíz, L. M. (2005). Adaptations to cave life in decapods from Oaxaca. *Association for Mexican Cave Studies Bulletin* 15, Austin: 170 pp.

Mejía-Ortíz L. M., Yañez G. and López-Mejía M. (2007). Echinoderms in an anchialine cave in Mexico. *Marine Ecology* 28(Suppl. 1): 31-36.

Mejía-Ortíz, L. M., Yañez G., López-Mejía M. and Zarza E. (2007). Cenotes from Cozumel Island, Quintana Roo, México. *Journal of Cave and Karst Studies*, 68(2) 250- 255.

Mejía L. M., Zarza, E. and López, M. (2008). *Barbouria yanezi sp. nov.*, A new species of cave shrimp (Decapoda, Barbouriidae) from Cozumel Island, Mexico. *Crustaceana*, 81(6): 663-672.

Mejía-Ortíz, L. M., Baldari, F. and López-Mejía, M. (2008). *Macrobrachium sbordonii* (Decapoda: Palaemonidae), A new stygobitic species of freshwater prawn from Chiapas, Mexico. *Zootaxa*, 1814: 49-57.

Mejía-Ortíz L. M. (2008). Adaptations from crustaceans to cave life. In: Rodríguez-Almaráz G. y F. Alvarez (Eds). *Crustaceans from Mexico; Currently Known*. Universidad de Nuevo León-Instituto de Biología UNAM.

Mejía-Ortíz, L. M., Yañez, G. and López-Mejía, M. (2017). Anchialocarididae, new family of anchialine decapod and a new species of genus *Agostocaris* from Cozumel Island, México. *Crustaceana*, 90(4): 381-398.

Mejía-Ortíz L. M. (2019). Crustacea. In: White, T., Culver D. and Pipan T. Eds. *Encyclopedia of Caves*. Wiley, London.

Reddell, J. R. (1981). A review of the cavernicole fauna of Mexico, Guatemala, and Belize. *Bulletin of the Texas Memorial Museum. The University of Texas at Austin*, 27: 1-327.

Rodríguez, G. (1992). The freshwater crabs of America. *Family Trichodactylidae and Supplement to the Family Pseudothelphusidae*. Editions d l'Orstom. Paris, 189 pp.

Rodríguez, G. and Smalley, A. E. (1969). The freshwater crabs from Mexico of the Pseudothelphusidae family (Crustacea, Brachyura). *Annals from Biology Institute*, UNAM, 40 Series Sea Sciences and Limnology, (1): 69-112.

Rodríguez, G., and Hobbs, H. H. Jr. (1989). Freshwater crabs associated with caves in southern Mexico and Belize, with description of three new species (Crustacea: Decapoda). *Proceeding of the Biological Society of Washington*, 102(2): 394-400.

Stenberg, v. R. and Shotte, M. (2004). A new anchialine shrimp of the genus *Procaris* (Crustacea: Decapoda: Procarididae) from the Yucatan

Peninsula. *Proceedings of the Biological Society of Washington*, 117(4): 514-522.

Villalobos, J. L., Alvarez, F. and Iliffe, T. M. (1999). New species of troglobitic shrimps from Mexico, with the description of *Troglomexicanus*, new genus (Decapoda: Palaemonidae). *Journal of Crustacean Biology*, 19(1): 111.122.

BIOGRAPHICAL SKETCH

Luis Manuel Mejia-Ortiz

Affiliation: University Of Quintana Roo.

Education: PhD. Marine Biology & Limnology University of Liverpool 2003.

MS Natural Aquatic Resources Management, National Autonomous University of Mexico, Marine Science & Limnology Institute 2001

Undergraduate in Biology, Metropolitan Autonomous University – Xochimilco, 1997.

Research and Professional Experience:
Evolutionary biology of cave crustaceans, Biodiversity of marine and freshwater crustaceans, Systematics of freshwater caridea (Palaemonidae); and Sustainable Development in Coastal and Inland Ecosystems

Professional Appointments: Researcher/Professor Full Time.

- Professor of PhD Sustainable Development
- Professor of Master in Sustainable Tourism Management
- Professor of the Undergraduate program Natural Resources Management

Publications from the Last 3 Years:

Baeza A. D. Hoilsten, Umaña-Castro R. & L. M. Mejía-Ortiz 2019. Population genetics and biophysical modeling inform metapopulation connectivity of the Caribbean king crab Maguimithrax spinosissimus, *Marine Ecology Progress Series*, 610:83-97.

Baeza, A., Umaña-Castro R. & L. M. Mejía-Ortiz 2019. Historical demography of the Caribbean spiny lobster *Panulirus argus* in the Florida Keys using genotype-by-sequencing derived SNPs *Journal of Crustacean Biology*. In Press.

Chavez-Solís, E. M., L. M. Mejía-Ortíz and N. Simoes. 2018. Predatory behavior of the cave shrimp Creaseria morleyi (Creaser, 1936) (Caridea: Palaemonidae), the blind hunter of the cenotes of Yucatán, Mexico. *Journal of Crustacean Biology*.

Cuadrado G., T. Kenkle & Mejía-Ortíz L. M. 2018. Regulatory challenges in realising integrated Coastal Management - Lessons from Germany, Mexico, Costa Rica and South Africa. *Sustainability,* 10, 3772.

Mejía-Ortíz L. M. 2019. Crustacea. In: White, Culver & Pipan Eds. *Encyclopedia of Caves*. Wiley, London. In Press.

Mejía-Ortíz L. M., M. López-Mejía, J. C. Tejeda-Mazariegos, O. Frausto-Martínez, K. A. Crandall, M. Pérez-Losada y J. Gaspar-Valladarez. 2018. Los camarones de agua dulce de la subfamilia Palaemoninae en la Península de Yucatán, (México, Guatemala y Belice). *Teoría y Praxis.* [Freshwater shrimp from the Palaemoninae subfamily in the Yucatan Peninsula, (Mexico, Guatemala and Belize). *Theory and Praxis.*]

Mejía-Ortíz, L. M. G. Yañez & M. López Mejía. 2017. Anchialocarididae, new family of anchialine decapod and a new species of genus *Agostocaris* from Cozumel Island, México. *Crustaceana*, 90(4): 381-398.

Mejía-Ortíz, L. M., T. Pipan, D. C. Culver and P. Sprouse. 2018. The blurred line between photic and aphotic environments: a large Mexican cave with almost no dark zone. *International Journal of Speleology*, 47(1): 1-12.

Tejeda-Mazariegos J. C., Mejía-Ortíz L. M, López-Mejía, M., Crandall, K. A., Pérez-Losada M. & Frausto-Martínez O. 2018. Freshwater crustaceans decapods an important resource of Guatemala. In: Sajal, R. *Biological Resources of Water* InTech Publisher. Serbia In press, 12 pp.

In: The Zoological Guide to Crustacea ISBN: 978-1-53616-366-7
Editor: Noelle Lachance © 2019 Nova Science Publishers, Inc.

Chapter 2

ARTEMIA FRANCISCANA (CRUSTACEA: ANOSTRACA) IN A HYPERSALINE HABITAT IN ABU DHABI (UNITED ARAB EMIRATES) AS INTERMEDIATE HOSTS FOR AVIAN CESTODES

Rolf K. Schuster[1,*], *Anitha Saji*[2]
and Shaika Salem Obaid Al Daheri[2]
[1]Central Veterinary Research Laboratory, Dubai, United Arab Emirates
[2]Environment Agency, Abu Dhabi, United Arab Emirates

ABSTRACT

Crustaceans of the genus *Artemia* inhabit hypersaline aquatic biotopes. They are economically important as their larval stages are used in aquaculture as food for fish and crayfish larvae. Serving as food source

---

[*] Corresponding Author's E-mail: r.schuster@cvrl.ae.

for waders and charadriform birds *Artemia* spp. play a role as intermediate host in the life cycle of avian cestodes.

The Al Wathba Wetland Reserve is a complex of natural and man-made water bodies about 40 km of central Abu Dhabi (United Arab Emirates). Founded in 1998 it has been recognized by the Convention of Wetlands of International Importance in 2013. In 2018, the wetland was listed in the IUCN Green List of Protected and Conserved Areas (UNEP-WCMC 2018). Apart from a stable population of greater flamingos more than 260 other bird species can be spotted there.

Between December 2017 and May 2018 a total of 2,700 *Artemia franciscana* collected at the Al Wathba Reserve were examined for the presence of larval helminths by direct microscopy of glycerin mounted specimens. Of these, 341 (= 12.6%) specimens contained cestode cysticercoids of eight) different species of the Hymenopepididae (*Flamingolepis liguloides*, *F. flamingo*, *Wardium fusa*, *Confluaria podicipina*), Dilepididae (*Eurycestus avoceti*, *Eurycestus* sp., *Anomotaenia tringae*) and Progynotaeniidae families (*Gynandrotaenia stammeri*). There was no significance in cysticercoid prevalence between male and female hosts. Flamingo specific species, *F. liguloides* and *F. flamingo*, were the most frequently found species. The number of cysticercoids varied between one and twelve and up to three species were detected concurrently in one shrimp. In addition, one shrimp contained a spirurid larval stage.

Keywords: *Artemia franciscana*, brine shrimps, cestodes, cysticercoids, Al Wathba Wetland Reserve, Abu Dhabi, United Arab Emirates

INTRODUCTION

Brine shrimps (*Artemia* spp.) is one genus within the the order Anostraca of the class Branchipoda. They are primitive, phylgenetically old microcrustaceans with a segmented body and leafe-like appendages (phyllopods). The genus inventory comprises seven morphologically similar bisexual species: *Artemia franciscana* Kellog, 1906, *A. monica* Verrill, 1869, *A. persimilis* Piccinelli & Prosdocimi, 1968, *A. salina* (Linnaeus, 1758), *A. sinica* Cai, 1989, *A. tibetiana* Abatzopoulos, Zhang & Sorgeloos, 1998 and *A. urmiana* Günther, 1899.

Two new species, *Artemia frameshifta* and *Artemia murae* (Naganawa & Mura, 2017), were recently found in Mongolia but their taxonomic status requires confirmation.

Various parthenogenetic populations of *Artemia* with different ploidy are combined as *A. parthenogenetica* but do not represent their own species.

Artemia spp. inhabit saline and hypersaline aquatic inland biotopes and can be found on all continents except Antarctica. The size of adult males and females is 8 -10 and 10 – 12 mm, respectively. Their life expectancy of *A. franciscana* depends on temperature, salinity and oxygen levels of the habitat (Browne & Wanigasekera 2000, Abatzopoulus et al. 2003). Female brine shrimps produce eggs that are ready to hatch almost immediately. Under extreme conditions (low oxygen, salinity above 15%) however, thick walled and metabolically inactive cysts are laidproduced. Due to their ability to tolerate high salt concentration of up to 25%, *Artemia* spp. exploit ecological niches that protect them from predators such as fish. The diet of brine shrimps consists of microscopic planktonic algae that are filtered from environment. Also, they are grazing on bottom mud and ingest helminth eggs.

Their role as intermediate hosts for cestodes was first disclosed for *Hymenolepis californicus* Young, 1950 by Young (1952). Intense research on brine shrimps was undertaken between the 1970s and the 1990s in Kazakhstan (Tengiz lake) and France (Camague) and later in different salters of the Mediterranean and Atlantic coasts of the Iberian Peninsula. As a result, it was ascertained that brine shrimps act as intermediate hosts for another 16 avian cestode species of the Hymenolepididae family (*Aploparaksis parafilum* Gasowska, 1932, *Brachiopodataenia gvozdevi* Maksimova, 1988, *Confluaria podicipina*, (Szymanski, 1905), *Fimbriarioides tadornae* Maksimova, 1976, *Flamingolepis caroli* (Parona, 1887), *F. flamingo* (Skrjabin, 1914) , *F. liguloides* (Gervais, 1847), *F. tengizi* Gvozdev & Maksimova, 1968, *Wardium fusca* (Krabbe, 1869) and *W. stellorae* Deblock, Biguet et Capron, 1960), the Dilepididae family (*Eurycestus avoceti* Clark,1954, *Anomolepis averini* Spassky & Yurpalova, 1967, *Anomotaenia tringae* (Burt,1940) and *A. microphallus* (Krabbe,

1869) and the Progynotaeniidae family *Gynandotaenia stammeri* Fuhrmann, 1936 and *Gynandrotaenia* sp. Redón et al., 2015 - a so far undetermined further species of this genus (Schuster 2018). The aim of this paper was to examine the occurrence of cestode cysticercoids in *A. franciscana* in the Al Wathba Wetland Reserve in Abu Dhabi, UAE.

METHODS

The Biotope

The Al Wathba Wetland Reserve is a complex of natural and man made surface water bodies situated 40 km southeast of central Abu Dhabi, the capital of the United Arab Emirates. The whole reserve streches over an area of 5 km^2 with the lake surface of approximately 1.6 km^2 (Saji et al. 2018) and became a protected wetland in 1998. Due to temporarily rising water levels resulting from sub-surface water flow and hydrostatic pressure from stronger winds and higher tides in winter (Dhaheri 2004) the place became a periodically natural wetland. A truck road that was constructed in the early 1980s dammed the flow of surface waters and increased the area and annual duration of surface waters. Treated waste water from a nearby Mafraq Water Treatment Plant transformed the wetland to a permanent water body that attracts a large variety of bird. Apart from Greater flamingos more than 260 bird species were spotted in the reserve (Sorae et al. 2014).

The brine shrimp, *A. franciscana*, is the main food source for the wading birds. The life cycle of the brine shrimps in the reserve begins when dorming cysts hatch when temperatures decline and salinity drops in autumn (Al Dhaheri & Saji 2013). Adult stages occur in November and reached highest concentration in May or June. They disappear due to high temperatures and rising salinity in summer leaving cysts behind.

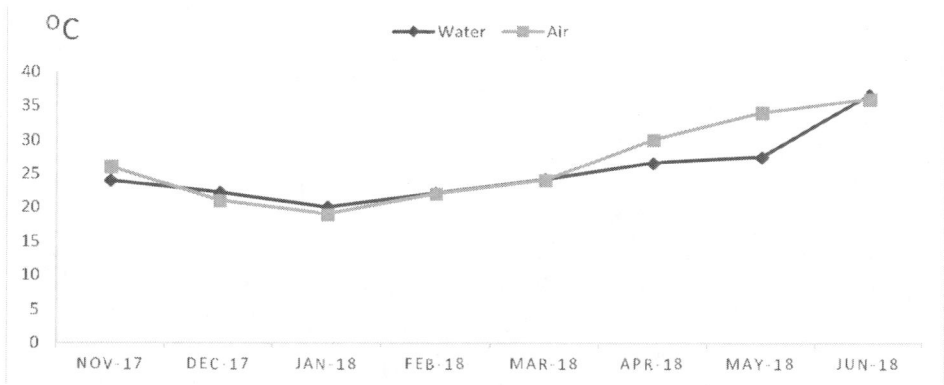

Figure 1. Average temperatures between November 2017 and June 2018. The water temperature is an average of individual measurements of all collection points at the Al Wathba Wetland Reserve. The air temperature was taken from Abu Dhabi International Airport Weather Station.

Collection and Examination of *A. fFranciscana*

In monthly intervals between December 2017 and May 2018, brine shrimps were collected by a 0.5 mm mesh sweep net. Up to 6 different points of the reserve were sampled, but not all the points could be reached all the time.

Examination of the crustaceans was carried out in the department of parasitology of the Central Veterinary Research Laboratory in Dubai. Shrimps were killed in boiling water and stored in 70% alcohol until examination. Randomly selected male and female specimens in numbers of 100 per site and collection date were put into a drop of glycerin and covered by a cover slip. Glycerin cleared the body content making cestode cysticercoids visible. Microscopic examination of the samples was carried out at low magnification (40 – 100x). Positive shrimps were dissected and isolated cysticercoids were studied in a drop of glycerin at higher magnification (400 – 600x). For species determination of the cestode larvae, the following parameters were used: colour, shape and size of the cysts, number shape and size of rostellar hooks and presence or absence of spines or crotchets on suckers. Measurements and photos were taken with an Olympus BX51 microscope connected to a camera Olympus DP 27 with the software Olympus cellSense dimensions. During the six-month period we examined a total of 2,700 adult specimens of *A. franciscana*.

Figure 2. The Al Wathba Wetland Reserve: Greater Flamingos with plovers, stilts and ruffs in the foreground in (Environment Agency Abu Dhabi).

Temperature Profile

The water temperature at all collecting points in the reserve was measured in monthly intervals when samples were taken (Figure 3). Additional measurements were made in the months before and after collection. The average monthly air temperature was obtained from from Abu Dhabi International Airport station which is at a distance of 22 km from Al Wathba Wetland Reserve (Annonymous 2019). (https://www.wunderground.com/history/monthly/ae/abu-dhabi/OMAA).

Statistical Treatment

Number and species composition of cysticercoids were noted in each individual male and female shrimp and were recorded in an Excel sheet. For calculating prevalence and intensity of infection at their 95% lower and upper confidence intervals the software package Quantitative Parasitology 3.0 (QP WEB) (Rozsa et al. 2000) was used.

RESULTS

A total of 341 out of 2,700 (=12.6%) were infected with eight different tapeworm larvae. However, the cysticercoid invasion was unequally distributed in the reserve. The lowest prevalence was found in sites 1, 2 and 7 (7.2, 10.0 and 10.2%, respectively). Site 10 had the higest number of infected shrimps bringing the prevalence to 18.6% (Table 1). The first sampling in December 2017 revealed 13% of examined shrimps infected (Table 2). The prevalence of infected shrimps dropped to 3% in January 2018 and increased in the February, March and April to 9, 14 and 21.7%, respectively. Itf fell at the end of observation in May to 8%. In June when water temperatures exceeded 35°C adult shrimps were not available anymore. There was no striking difference in the percentage of infected male (12.1%) and female (13.3%) *A. franciscana*.

F. liguloides was the most frequent cestode larval stage. It occurred in numbers between one and eight in a total of 242 shrimps. *F. liguloides* mono infections were seen in 221 hosts. Twenty-one further shrimps harbored mixed infections with *F. liguloides* with two or three other cestode species (Table 3). *F. flamingo* was counted in 60 shrimps in numbers between one and twelve. The species occurred in 40 cases as mono infection and in 20 further shrimps in combination with two or three other other species. Thirty-five crustaceans were infected with cysticercoids of the genus *Eurycestus*. The majority of *E. avoceti* cysticercoids was found as mono infections 21 shrimps, and only nine were seen in combination with other species. Only single specimens of a further species of this genus, *Eurycestus* sp., were found in five crustacean hosts. Cysticercoids of *W. fusa*, *A. tringae*, *G. stammeri* and *C. podicipina* were detected in fifteen, ten, six and five shrimps, respectively. The number of *A. franciscana* infected with cysticercoids of different species is shown in Table 4.

One artemia contained a single spirurid larval stage.

Table 1. Prevalence and intensity and their 95% confidence intervals of cysticercoids in Artemia franciscana collected at different sites of the Al Wathba Wetland Reserve in Abu Dhabi, UAE

Collection site	Number examined	infected	Prevalence %	95% C.I.	Intensity average	95% C.I.
1	400	29	7.2	4.9;10.2	1.38	1.14;1.66
2	400	40	10.0	7.2;13.4	1.45	1.18;2.02
4	400	61	15.20	11.9;19.2	1.51	1.32;1.69
6	500	67	13.4	10.5;16.7	1.21	1.10;1.33
7	500	51	10.2	7.7;13.2	1.47	1.20;2.37
10	500	93	18.6	15.3;22.3	1.44	1.29;1.65
total	2,700	341	12.6	11.4;13.9	1.41	1.33;1.54

Table 2. Prevalence and intensity and their 95% confidence intervals of cysticercoids in *Artemia franciscana* collected between December 2017 and May 2018 in the Al Wathba Wetland Reserve in Abu Dhabi, UAE

Collection month	sites	Number examined	infected	Prevalence %	95% C.I.	Intensity average	95% C.I.
December	2,6,10	300	39	13.0	9.4;17.3	1.44	1.23;1.70
January	1,4,7,	300	9	3.0	1.4;5.6	1.44	1.00;2.33
February	1,2,4,6,7,10	600	54	9.0	6.8;11.6	1.31	1.17;1.48
March	1,2,4,6,7,10	600	84	14.0	11.3;17	1.30	1.13;1.83
April	1,2,4,6,7,10	600	130	21.7	18.4;25.2	1.57	1.42;1.76
May	6,7,10	300	25	8.3	5.5;12.1	1.08	1.00;1.20

Table 3. Species combination of cysticercoids in *Artemia franciscana* collected between December 2017 and May 2018 in the Al Wathba Wetland Reserve in Abu Dhabi, UAE

Species	Number of cestode species with multiple infections												
	2										3		
F. liguloides	+	+	+	+	+						+	+	
F. flamingo	+					+	+	+	+	+	+	+	
C. podicipina		+			+								
W. fusca						+					+		
E. avoceti			+			+		+	+			+	
A. tringae				+			+	+					
G. stammeri				+			+		+				
N infected:	10	1	5	2	1	1	3	1	1	2	1	1	1	1

Figure 3. Collecting sites of *Artemia franciscana* in the Al Wathba Wetland Reserve.

Table 4. Number of *Artemia franciscana* infected with cysticercoids of different species in the Al Wathba Wetland Reserve between December 2017 and May 2018

Month	*F. liguloides*	*F. flamingo*	*W. fusa*	*C. podicipina*	*E. avoceti*	*E.* sp.	*A. tringae*	*G. stammeri*
Dec	7	15	13	5	3		2	5
Jan	3	3			3			
Feb	44	6	2		3		6	
Mar	58	10			14	3		1
Apr	106	25			7	2		
May	24	1						
Total:	242	60	15	5	30	5	8	6

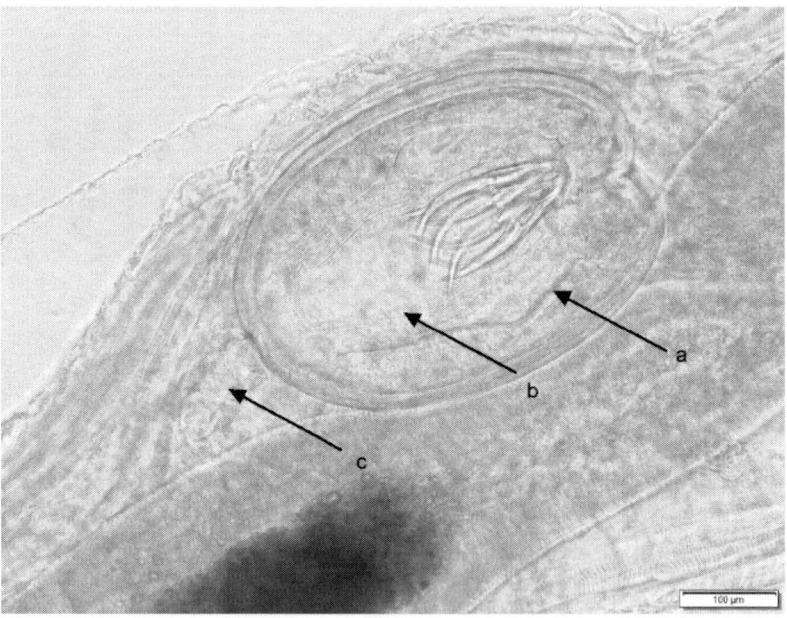

Figure 4. *Flamingolepis liguloides* cysticercoid in the abdomen of *A. franciscana*. The relatively big cysticercoid has rostellar hooks of around 180 µm. a: sucker, b: rostellum, c: cercomer.

Artemia Franciscana (Crustacea: Anostraca) ... 77

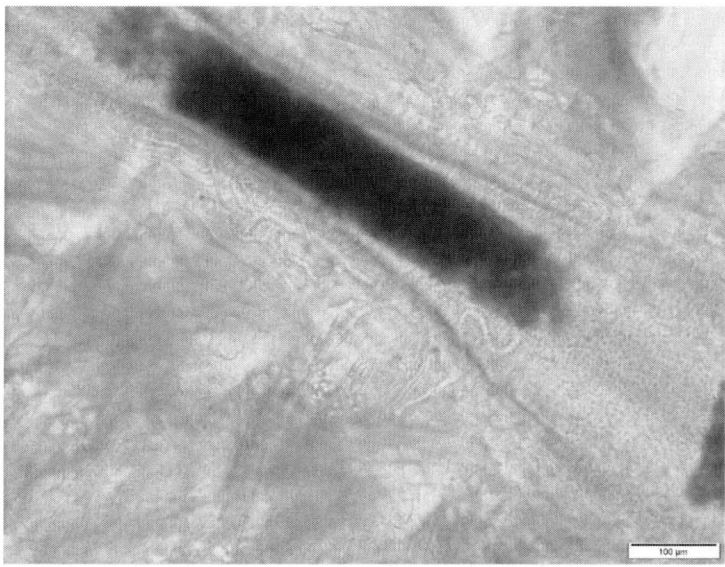

Figure 5. *Flamingolepis flamingo* cysticercoid with a long and much convoluted cercomer in the thorax of *A. franciscana*.

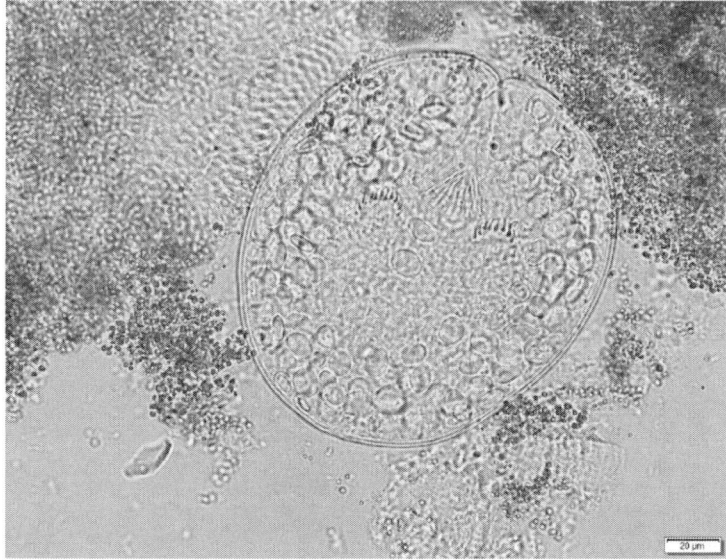

Figure 6. *Eurycestus avoceti* cysticercoid freed from surrounding brown capsule. The anterior rims of suckers are armed with 12 to 14 crochets.

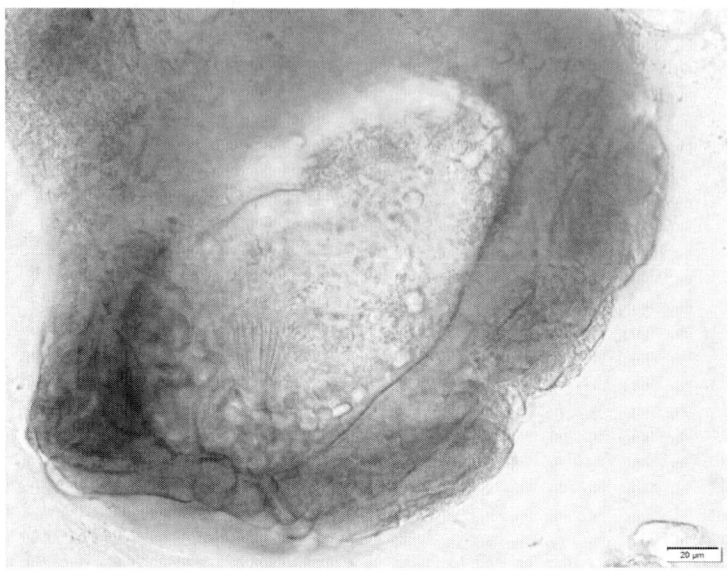

Figure 7. *Eurycestus* sp. cysticercoid surrounded by a flaky brown capsule. Contrary to *E. avoceti* this cysticercoid is elongated and suckers are armed with a larger number of crochets.

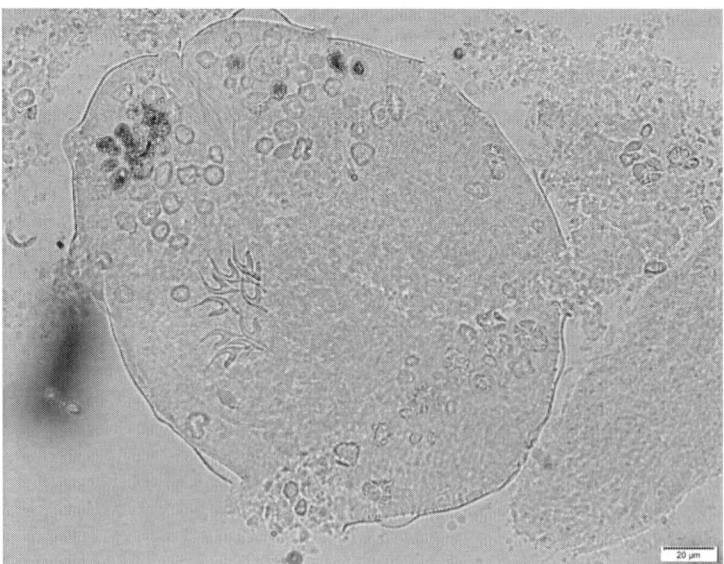

Figure 8. *Wardium fusa* cysticercoid. The cysticercoid was squeezed to observe and measure rostellar hooks.

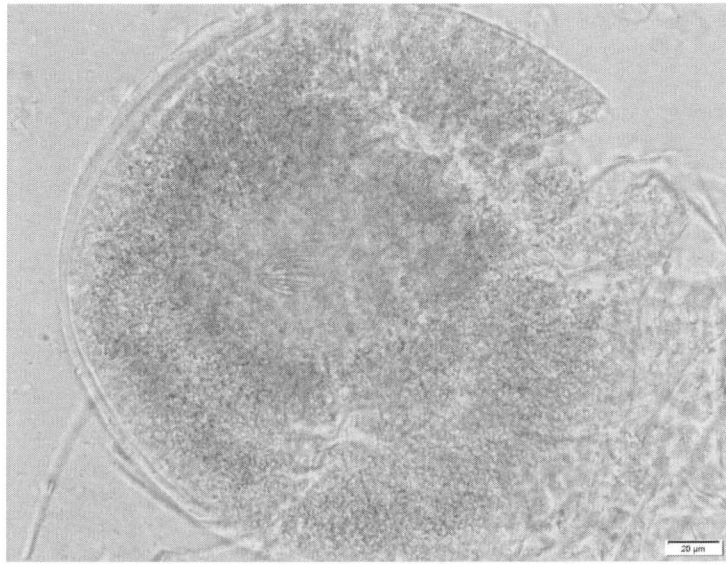

Figure 9. The *Anomotaenia tringae* cysticercoid surrounded by an amorphous brown capsule.

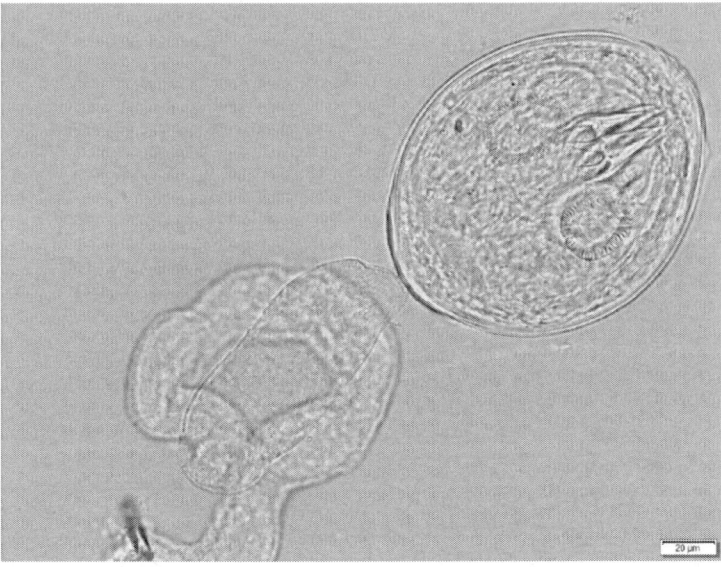

Figure 10. Isolated *Gynandrotaenia stammeri* cysticercoid with a long cercomer and suckers armed with densely arranged spines.

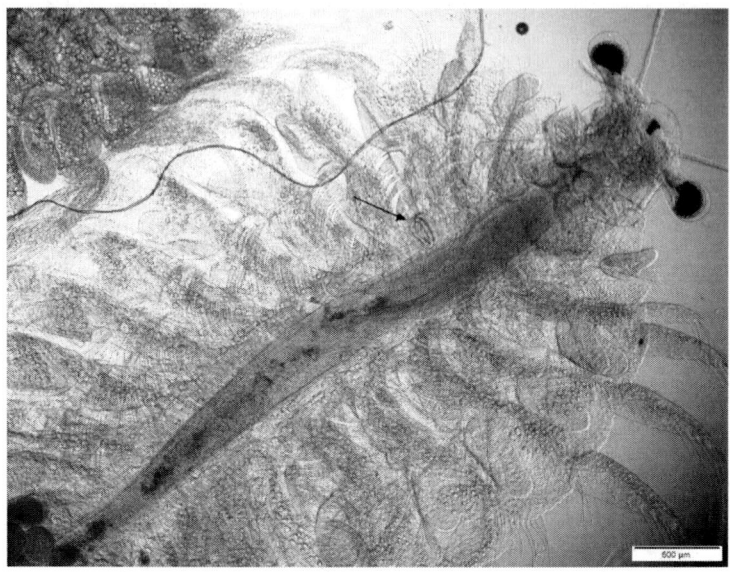

Figure 11. Spirurid larva at the base of a phyllopode in a female *A. franciscana*.

DISCUSSION

The source of the brine shrimps in the Al Wathba Reserve is unknown and it is suggested that cysts had been introduced with migrating birds (Al Dhaheri 2004). The only other known place where brine shrimps occur in the UAE are the Godolphin lakes in Dubai. There, *A. franciscana* was introduced in 1998 as food source for flamingoes and other waders and shore birds. The species affiliation of the brine shrimps to *A. franciscana* in the Al Wathba Reserve was confirmed in a molecular study by Saji et al. (2019).

The highest percentage of infected shrimps were found at site No. 10. This site is closest to the inlet of treated waste water from Mafraq Sewage Treatment Plant. With an average of 100 ppt it had the lowest salinity and showed the highest artemia density (Al Dhaheri & Saji 2013) and attracted many birds. Monthly differences in the general cysticercoid prevalence (Table 2) can be explained by the profile of water temperature Figure 31). The first generation of brine shrimps that hatched from cysts after the hot summer in November 2017 benefited from ambient water temperatures of 24°C. Increased activity of the crustaceans necessitated a high food uptake including ingestion of cestode eggs. This resulted in a relative high cysticercoid prevalence in December 2017. Attributed to a life span of two to three months ((Browne & Wanigasekera 2000), members of this generation faded in coming months. Lower temperatures of 20°C in January decreased the activity of the crustaceans and vanishing of infected specimens caused a drop in cysticercoid prevalence to 3%. As a result of rising water temperatures in the following months, the percentage of infected shrimps rose again and reached maximum of 223% in April. Since migrating birds and other temporary visitors had left the habitat in March and April, contamination of the water with cestode eggs diminished. This resulted in a decreased cysticercoid prevalence in May. With rising temperqatures above 24°C the life expectancy of *A. franciscana* decreases drastically to less than four weeks. And for this reason, also heat stress leading to a die off of infected specimens might have contributed to the drop in prevalence. Brine shrimps faded when water temperature rose to 36°C in June.

The genus *Flamingolepis* was erected by Spassky and Spasskaya (1954) to accommodate relatively small hymenolepidid cestodes with a rostellum bearing eight skrjabinoid hooks. Final hosts are flamingos. The species inventory consists of a total of seven names: *F. caroli* (Parona, 1887), *F. chileno* Babero et al, 1981, *F. dolgushini* Gvozdev & Maksimova, 1968, *F. flamingo* (Skrjabin, 1914), *F. liguloides* (Gervais, 1847), *F. megalorchis* (Lühe, 1898) and *F. tengizi* Gvozdev & Maksimova, 1968.

Although there is confusion about the validity of the species of the genus *Flamingolepis*, in this publication we follow the suggestion Robert & Gabrion (1991) and recognize large cysticercoids with eight skrjabinoid hooks in a range of 180 µm in legth as larval stage of *F. liguloides*[1].

---

[1] The original description by Gervais (1847) was superficial and rostellar hooks were not mentioned. In a redescription of *F. liguloides,* Lühe (1898) gave hook lengths of 130 µm. Most probably Lühe mixed up *F. liguloides* with *F. caroli* that had been described earlier from flamingos in Sardinia by Parona (1887).

Cysticercoids of two representatives of the genus *Flamingolepis*, *F. liguloides* and *F. flamingo*, were the most frequently recorded species in *A. franciscana* in the Al Wathba reserve. They have been previously also reported from *A. franciscana* from an artificial hypersaline pond in Dubai (Sivakumar et al. 2018) and were found in *A. franciscana*, *A. salina* and *A. parthenogenetica* populations in salinas in the western Mediterranean and Atlantic coasts (Robert & Gabrion 1991, Georgiev et al. 2005, 2007, Sanchez et al. 2007, Redon et al. 2011, Sanchez et al. 2013, Georgiev et al. 2014 and others). Both cestode larval stages were also described from *A. salina* in the lake Tengiz in Kazakhstan under the junior synonyms *F. dolgushini* and *F. tengizi*, respectively. Both species are obligate intestinal parasites of flamingos.

Cysticercoids of the genus *Eurycestus* were presented with two species, *E. avoceti* and a so far not named *Eurycestus* sp. Yet, three different species, namely *E. avoceti* Clark, 1954, *E. falciformis* Burt, 1979 and *E. latissimus* Burt, 1979, had been described as adult cestodes from American avocets (*Recurvirostra americana* Gmelin, 1789) in the USA. The original descriptions were based on mature strobilae without scolex. Baer (1968) completed the description of *E. avoceti* by characterizing the whole cestode including scolex with material from pied avocette (*R. avosetta* Linneus, 1758) from French Camargue. Apart from avocets, greater flamingos, black winged stilts (*Himantopus himantopus* Linneus, 1758) and slender billed gulls (*Chroicephalus genei* Breme, 1839) were mentioned as final hosts for *E. avoceti* in a paper by Maksimova (1991). *Eurycestus* cysticercoids occurred in brine shrimps in different hypersaline wetlands of France, Spain and Portugal in prevalences between 0.09 and 30.0% and were also detected in in the Godolphin lakes of Dubai (Schuster 2019). Cysticercoids of both species differ in the structure of the surrounding capsule, the shape of the cysticercoid and the number of crochets on suckers.

Fifteen shrimps were infected with single cysticercoids of *Wardium fusa* during winter months. *W. fusa* is a tapeworm of gulls. Maksimova (1987) found the adult stages of the tapeworm in the intestines of black headed gulls, European herring gulls, common gulls, 1758 and Pallas's gulls caught at the Tengiz lake in Kazakhstan and described the subsequent cysticercoid in *A. salina*. Gulls of three species (black headed gulls, slender-billed gulls and lesser black-billed gulls) were often seen between in great flocks in the Al Wathba reserve from November to March. This would explain why cysticercoids were found only during winter months. In previous publications about tapeworm larvae in brine shrimps in western Europe researchers (Robert & Gabrion 1991, Georgiev et al. 2005, 2007, Sanchez et al. 2007, 2013, Redon et al. 2015) described *W. stellorae*, a species that differs from *W. fusa* in larger dimensions of rostellar hooks (20-22 vs. 15-18 µm).

Cysticercoids of *Anomotaenia tringae* were found in eight *A. franciscana*. The adult cestodes parasitize sandpipers of the family Scolopacidae Rafinesque, 1815 that includes shenks and tattlers, and also in other waders. These birds are seen in large numbers in the Al Wathba reserve during winter months. Most of these species are long distance migratory birds. Apart from *A. tringae*, Georgiev et al. 2005 described another cysticercoid, *Anomotaenia* sp., with 26 to 30 rostellar hooks in two rows of a length of 12-13 and 11-12 µm, rescectively while *A tringae* processes 18 or 20 hooks with a length of 19 -21 µm.

Six brine shrimps were infected with *Gynandrotaenia stammeri*. The cysticercoid with armed suckers, six rostellar hooks and a long coiled cercomer was first described by Gvozdev & Maksimova (1979) in *A. salina* from Tengiz lake. It was also found in brine shrimps in the French Camargue by Robert & Gabrion (1991), in the Odil marshes of Spain (Georgiev et al. 2005, Sanchez et al. 2007, Sanchez et al. 2013) and in salinas of the Atlantic coasts of Portugal and Spain (Georgiev et al. 2007). The adult worm was described by Fuhrmann (1936) from material found in a flamingo in Poland (Stammer 1935).

Our finding of a spirurid larva confirmed also the role of *Artemia* spp. as intermediate host for nematodes. A large number of spirurid nematodes parasitize wader birds (Barus et al. 1978) and representatives of the nematode family Tetrameridae were found in the proventriculus of flamingos (Lühe 1898, Gvozdev & Maksimova 1971).

In summarizing the results, it can be concluded that *A. franciscana* is not only a food source for birds residing or visiting the Al Wathba Wetland Reserve, these brine shrimps are also a source of helminth infection and detected cysticercoids reflect the permanent or temporary presence of avian host species in the habitat.

ACKNOWLEDGMENTS

The authors are greatefull to Prof. Dr. D. Erhan (Institut of Zoology Chisinau/ Republik of Moldova) to Prof. Dr. V. KCharchenko (Schmalhausen Institue of Zoology, Kiev/ Ukraine) and to Prof. Dr. S. Rehbein (Boehringer-Ingelheim, Rohrdorf/ Germany) for their help in acquisition of historical literature sources.

REFERENCES

Abatzopoulos, T. J., El-Bermawi, N., Vasdekis, C., Athanasios D. Baxevanis, A. D., Sorgeloos, P. (2003). Effects of salinity and temperature on reproductive and life span characteristics of clonal *Artemia*. (International Study on *Artemia*. LXVI). *Hydrobiologia* 492: 191-199.

Al Dhaheri S. S. (2004). *Assessment of brine shrimp (Artemia sp.) Productivity at Al Wathba wetland reserve, Abu Dhabi (UAE)*. BSc Thesis, United Arab Emirates University, Abu Dhabi, UAE. 148 pp.

Al Dhaheri, S., Saji, A. (2013). Water quality and brine shrimp (*Artemia* sp.) population in Al Wathba Lake, Al Wathba Wetland Reserve, Abu Dhabi Emirate, UAE. *International Journal of Biodiversity and Conservation*. 5: 281-288.

Annonymous 2019. *Temperature profile for Abu Dhabi* (Nov. 2017-Jun. 2018). https://www.wunderground.com/history/monthly/ae/abu-dhabi/ OMAA.

Baer, J. G. (1968). *Eurycestus avoceti* Clark, 1954 (Cestode cyclophyllidien) parasite de l' avocette en Camargue. *Vie et Milieu,* 19: 189-198. [*Eurycestus avoceti* Clark, 1954 (Cestoda; Cyclophyllidea) a parasite of the avocet in the Camargue]. *Vie et Milieu,* 19: 189-198].

Barus, V., Sergeeva, T. P., Sonin, M. D., Ryzikov, K. M. (1978). *Helminths of fish-eating birds of the Palaearctic region 1. Nematoda.* Academica, Publishing house of the Czechoslovak Academy of Sciences, Praha. 319 pp.

Browne, R. A., G. Wanigasekera, G. (2000). Combined effects of salinity and temperature on survival and reproduction of five species of *Artemia. Journal of Experimental Marine Biology and Ecology* 244: 29–44.

Georgiev, B. B., Angelov, A., Vasilieva, G. P., Sanchev, M. I., Hortas, F., Mutavchiev, Y., Pankov, P., Green, A. J. (2014). Larval helminths in the invasive American brine shrimp *Artemia franciscana* throughout its annual cycle. *Acta Parasitologica* 59: 380-389.

Georgiev, B. B., Sanchez, M. I., Green, A. J., Nikolov, P. N., Vasilieva, G. P., Mavrodieva, R. S. (2005). Cestodes from *Artemia parthenogenetica* (Crustacea, Branchipoda) in the Odiel Marshes, Spain: a systematic survey of cysticercoids. *Acta Parasitologica* 50 (2): 105-117.

Georgiev, B. B., Sanchez M. I., Nikolov, P. N. Vasilieva, G. P., Green, A. J. (2007): Cestode parasitism in invasive and native brine shrimps (*Artemia* spp.) as a possible factor promoting the rapid invasion of *A. franciscana* in the Mediterranian region. *Parasitology Research* 101: 1664-1655.

Gervais P. (1847). Sur quelques entozoires taenoides et hydatides. *Academie des Sciences et Lettres de Montpellier. Memoires de la section des sciences* 1: 85-98. [On entozooans taenoids and hydatids. *Academy of Sciences and Letters of Montpellier. Memories of the Science Section* 1: 85-98].

Gvozdev, E. V., Maksimova, A. P. (1971). K gel'mintofaune rozovogo flamingo (*Phoenicopterus roseus* Pall.) v Kazachstane. *Trudy Instituta Zoologii* 31: 41-46. [On the helminth fauna of greatyer flamingos

(*Phoenicopterus roseus* Pall.) *Proceedings of the Institute of Zoology* 31: 41-46].

Fuhrmann O. (1936). Gynandrotaenia stammeri nov. gen nov. spec. *Revue Suisse de Zoologie* 43: 517-518. [*Swiss Journal of Zoology*]

Lühe, M. (1898). Beiträge zur Helminthenfauna der Berberei. *Sitzungsberichte der Königlich Preussischen Akademie der Wissenschaften zu Berlin.* (2): 618-628. [Contributions on the helminth fauna of the Barbary. *Session reports of the Royal Preussian Academy of Sciences at Berlin.* (2): 618-628].

Maksimova, A. P. (1987). K morfologii I ciklu razvitija custody *Wardium fusa* (Cestoda, Hymenolepidodae). *Parazitologyia* 21: 157-159. [On morphology and developmental cycle of *Wardium fusa* (Cestoda, Hymenolepidodae). *Parazitologyia* 21: 157-159].

Maksimova, A. P. (1991). К экологии и биологии *Eurycestus avoceti* (Cestoda: Dilepididae). Parazitologiya 25: 73-76. [On ecology ad biology of *Eurycestus avoceti* (Cestoda: Dilepididae). *Parazitologiya* 25: 73-76].

Naganawa, H., and Mura, G. (2017). Two new cryptic species of *Artemia* (Branchiopoda, Anostraca) from Mongolia and the possibility of invasion and disturbance by the aquaculture industry in East Asia. *Crustaceana* 90, 1679–1698.

Parona, C. (1887). Elmintologia Sarda. Contribuzione allo Studio dei vermi parassiti in animali di Sardegna. *Annali de Museo civico di storia natural di Genova.* IV: 301-303. [Helminthology of Sardinia. Contribution to the study of worm parasites in animals of Sardinia. *Annals of the Civic Museum of Natural History of Genoa.* IV: 301-303].

Redon, S., Amat, F., Hontoria, F., Vasilieva, G., Nikolov, P. N., Georgiev, B. B. (2011). Participation of metanaulplii and juvenile individuals of *Artemia parthenogenetica* (Branchiopoda) in the circulation of avian cestodes. *Parasitology Re*search 108: 905-912.

Redon, S., Green, A., Georgiev, B. B., Vasilieva G. P. Amat, F. (2015). Influence of development stage and sex on infection of the American brine shrimp *Artemia franciscana* Kellog, 1906 by avian cestodesin Ebro Delta salterns, Spain. *Aquatic Invasions* 10: 415-423.

Redón, S., Green, A. J., Georgiev, B. B., Vasileva, G. P., Amat, F. (2015). Influence of developmental stage and sex on infection of the American brine shrimp *Artemia franciscana* Kellogg, 1906 by avian cestodes in Ebro Delta salterns, Spain. *Aquatic Invasions* 10: 415-423.

Robert, F., Gabrion, C (1991). Cestodes de l'avifaune Camarguaise. Role d' *Artemia* (Crustacea, Anostraca) et strategies de rencontre hote-parasite. *Annales de Parasitology Humaine et Comparee* 66: 226-235. [Cestodes of birds in Camargue. Importance of *Artemia* (Crustacea, Anostraca) and meeting strategie between hosts and parasites. *Annals of Human and Comparative Parasitology* 66: 226-235].

Rozsa, L., Reiczigel, J., Majoros, G. (2000). Quantifying parasites in samples of hosts. *Journal of Parasitology* 86: 228-232.

Saji, A., Eimanifar, A., Soorae, P. S. Al Dhaheri, S., Li, W., Wang, P. Z., Asem, A. (2019). Phylogenetic analysis of exotic invasive species of the brine shrimp *Artemia* Leach, 1819 (Branchiopoda, Anostraca) in Al Wathba Wetland Reserve (U.A.E.; Abu Dhabi). *Crustaceana* 92: 495-503.

Saji, A., Mischke, S. Sorae, P. S., Ahmed, S., Al Dhaheri, S. (2018). The Al Wathba Wetland Reserve Lake in Abu Dhabi, United Arab Emirates and its ostracod (seed shrimp) fauna. *International Journal of Aquatic Biology.* 6: 265-273.

Sanchez M. I., Georgiev, B. B., Green, A. J. (2007). Avian cestodes affect the behavior of their intermediate host *Artemia parthenogenetica*: An experimental study. *Behavioural Processes* 74: 293-299.

Sanchez M. I., Nikolov, P. N., Georgieva, D. D., Georgiev, B. B., Vasilieva, G. P., Pankov, P., Paracuellos, M., Lafferty K. D., Green, A. J. (2013). High prevalence of cestodes in *Artemia* spp. throughout the annual cycle: relationship with abundance of avian final hosts. *Parasitology Research* 112: 1913-1923.

Schuster, R. K. (2018). The role of *Artemia* spp. (Branchiopoda: Artemiidae) as intermediate hosts for avian cestodes. *Environmental Analysis and Ecological Studies* 1(4). EAES.000518.2018.

Schuster, R. K. (2019). On two morphologically different cysticercoids of the genus *Eurycestus* (Cestoda: Dilepididae) in *Artemia franciscana* (Arthropoda: Artemiidae) in a hypersaline pond in Dubai, United Arab Emirates. *Helminthologia* 56: 151-156.

Sivakumar, S. Hyland, K., Schuster, R. K. (2018). Tapeworm larvae in *Artemia franciscana* (Crustacea: Anostraca) in the Godolphin lakes of Dubai (United Arab Emirates) throughout an annual cycle. *Journal of Helminthology*. doi:10.1017/S0022149X18000913.

Soorae, P. S., Al Zaabi, R., Saji, A., Sakkir, S., Ahmed, S., Khan S. (2014). *Al Wathba wetland reserve field guide*. Environment Agency – Abu Dhabi, UAE. 84 pp.

Spassky, A. A., Spasskaja, L. P. (1954). O postroenii sistemy gimenolepidid. *Trudy Gel'mintologiceskoj Laboratorii Akademii Nauk SSSR* 7: 55-119. [On the construction of a system of Hymenolepididae. *Proceedings of the Gel'mintological Laboratory of the USSR Academy of Sciences* 7: 55-119].

Stammer, H. J. (1935). Die Entoparasiten der in Schlesien 1935 beobachteten Flamingos. *Berichte des Vereins Schlesischer Ornithologen*. 21: 15-17. [Endoparasites of flamingos sighted in Silesia in 1935. *Reports of the Association of Silesian Ornithologists*. 21: 15-17].

UNEP-WCMC (2018). *Protected Area Profile for Al Wathba from the World Database of Protected Areas*, December 2018. Available at: www.protectedplanet.net.

Young, R. T. (1952). The larva of *Hymenolepis californicus* in the brine shrimp (*Artemia salina*). *Journal of Washington Academy of Science* 42, 385-388.

In: The Zoological Guide to Crustacea ISBN: 978-1-53616-366-7
Editor: Noelle Lachance © 2019 Nova Science Publishers, Inc.

Chapter 3

BEHAVIOURAL RESPONSES OF THE NON-MARINE OSTRACOD *HETEROCYPRIS IINCONGRUENS* (CRUSTACEA: OSTRACODA) TO NATURAL CHEMICAL CUES: LABORATORY OBSERVATIONS

Dragana Miličić[1,*], *Tatjana Savić*[2], *Branka Petković*[2], *Milica Potrebić*[1], *Jelena Trajković*[1] *and Sofija Pavković-Lučić*[1]

[1]University of Belgrade--Faculty of Biology, Belgrade, Serbia
[2]Institute for Biological Research "Siniša Stanković"
Institute for Biological Research,
University of Belgrade, Belgrade, Serbia

---

[*] Corresponding Author's E-mail: draganam@bio.bg.ac.rs.

ABSTRACT

In aquatic environments, chemical cues are very important for the perception of danger, especially when visibility is low. It is known that ostracods rely on chemical senses to detect predators, which is essential for survival. Occurrence of alarm signals in the surroundings can affect their activities and behaviour. The present study investigates if the non-marine ostracod *Heterocypris incongruens* (Crustacea: Ostracoda) can detect and react to chemical compounds derived from a predator (*Triturus* spp. larvae) and from injured conspecifics. Also, the study aims to investigate whether habitat-substrate selection is at play when the individuals are in potential danger. The obtained results indicate that predator-derived chemicals and conspecific alarm cues induce specific behavioural responses: forming of aggregations, reduction of locomotion or camouflaging. It is possible that *H. incongruens* rapidly evaluates environmental cues and modifies defensive strategies depending on the type of semiochemicals perceived.

Keywords: *Hheterocypris incongruens*, chemical cues, antipredator behaviour

INTRODUCTION

Predation represents a selective force shaping ecological and evolutionary dynamics of prey populations (Tollrian and Harvell, 1999). During the evolutionary arms race in predator-prey interactions, myriad mechanisms reducing mortality by predation have evolved in prey. Antipredator responses can include changes in morphological, behavioural and life-history traits (Endler, 1995). However, probably the most rapid and most commonly expressed group of defence mechanisms is behavioural plasticity (Hazlett, 2011). Basically, two forms of antipredator behaviour strategies can be distinguished: 1) escaping detection by the predators and 2) forms of behaviour manifested when prey confronts a potential predator (Dugatkin, 2013). Behavioural defence strategies may include mimicry, crypsis, thanatosis, altered habitat use, shelter seeking, reduction of foraging activity, etc. (Ferrari et al., 2010; Mitchell et al.,

2017). Whatever the case, early detection of the predator represents a key part of any defensive behavioural strategy (Edmunds, 1974; Ferrari et al., 2010).

In aquatic environments, dissolvable and a rapidly-spreading chemical cues are essential for perception of a predator's presence, anticipation of possible attack and devising a successful defensive response (Barreto et al., 2013). These cues are especially important when visibility is low, since they provide valuable information about the spatial and temporal distribution of predation risks (Mitchell et al., 2017). The predator-prey chemical communication phenomenon involves extremely diverse groups of infochemicals (Lass and Spaak, 2003). These compounds can be generally divided into three groups: chemicals emitted by the predators – kairomones (i.e., predator odour); startled prey-derived compounds – disturbance cues (i.e., pulsed release of urinary ammonia) and injured prey-derived chemicals – alarm cues (i.e., blood) (Ferrari et al., 2010). Each of the listed types of chemical compounds induces a variety of different behavioural responses that can be species-specific and can vary even among individuals of the same taxa (Templeton and Shriner, 2004; Wilson et al., 2005; Lakowitz et al., 2008; Orr et al., 2009).

Crustaceans, when exposed to kairomones, tend to reduce the levels of their activity, change their position in the water column, display grouping behaviour or exploit a structural refuge [reviewed in James and McClintock (2017)]. Among crustaceans, non-marine ostracods represent a valuable group of model systems (Holmes and Chivas, 2002; Külköylüoğlu et al., 2010) used for behavioural assays exploring feeding (Vannier et al., 1998; Schmit et al., 2007; Miličić et al., 2015) or defensive strategies and responses to different stress factors (Fernandez et al., 2016). Although fossils of their carapaces have been found from the Lower Ordovician (Williams et al., 2008), the group can be traced back to the mid-to-late Cambrian (Harvey et al., 2012). The non-marine ostracods have a wide distribution, ranging from low latitudes to Arctic water bodies (Delorme, 1991; Külköylüoğlu et al., 2010). These crustaceans are members of benthic and littoral communities (Delorme, 1991; Fryer, 1997; Mesquita-

Joanes et al., 2012) and are considered highly tolerant to different environmental stress factors, especially drought (Horne, 1993).

Rossi et al. (2013) and Vandekerkhove et al. (2013) explored traits conferring on ostracods the ability to persist in some unique aquatic systems (such as shallow seasonal pools and temporary ponds). The authors identified specific features of *Heterocypris incongruens* and *Eucypris virens* that contribute to their tolerance to hypoxia, high salinity and UV radiation. Many ostracods are good swimmers, while many others crawl and burrow through the substrate.

Heterocypris incongruens belongs to the order Podocopida. Members of the order inhabit continental waters and have almost cosmopolitan distribution. Relationships with other species can contribute to this wide distribution (Fernandez et al., 2016), and one of these inter-specific relationships may involve predator-prey interactions. Various fish species (usually absent in temporary habitats), aquatic insects (especially Odonata larvae) and amphibians have been recorded consuming *H. incongruens* (Havel et al., 1993; Johnson, 1995; Ottonello and Romano, 2011; Vandekerkhove et al., 2012). Other crustaceans, such as decapod, copepod and amphipod species, also prey on ostracods (Moguilevsky and Gooday, 1977). On the other hand, *H. incongruens* occasionally has been observed attacking some macroinvertebrates, such as cladocerans, copepods, chironomids and oligochaetes (Liperovskaya, 1948; Ganning, 1971). Larger prey can also be on the menu, especially frog eggs and sickly or injured animals (fish, amphibians, even water birds) (Reichholf, 1983; Ottonello and Romano, 2011). In addition, cannibalism has also been reported in clonal females of *H. incongruens* (Rossi et al., 2011). They consumed soft body parts of conspecific specimens that released their body fluids into the surrounding medium, causing even more aggressive behaviour by the ostracod swarms. Scavengers usually attack in groups and are able to ingest massive quantities of the prey's tissue (Wilkinson et al., 2007).

Ostracods rely on chemical senses to locate food sources or detect danger and predators, which is a feature essential for survival. The occurrence of alarm signals in the surroundings can be a source of stress, and such rather "uncomfortable" situations will affect the behaviour of individuals. The present study deals with the responses of *H. incongruens* to such signals, even when the predator or predator risk is not visually identified. The applied design mimics two situations in nature: 1) the presence of chemical compounds derived from the predator (kairomones); and 2) the presence of alarm cues released from injured conspecific

individuals. Another aim of the study was to determine whether specific habitat-substrate selection is at play when individuals are in danger, a goal achieved by the use of dark- and light-coloured substrates.

MATERIAL AND METHODS

Animals and Chemical Compounds

Individuals of *H. incongruens* were sampled in a natural pond in eastern Serbia (43.0623° N 22.6946° E) during the summer of 2014 and maintained under laboratory conditions in aquarium water with detritus from the pond of origin. Two weeks prior to the experiment, samples were kept at room temperature (20 ± 2°C) under a light regime consisting of 16 h of light and 8 h darkness. They were fed weekly with dry *Spirulina* condensed commercial food. All specimens used in the experiment (Figure 1A) were in the adult stage (with a size of 1.5 mm) and belonged to a single population.

In natural conditions, *Triturus* larvae are predators feeding on various small invertebrates, including ostracods (Fasola and Canova, 1992). Larvae of the newt *Triturus* spp. (Caudata) hatched in the laboratory (Figure 1B) were used to obtain "water with predator" chemical cues.

From the moment the experimental larvae formed a mouthpiece, they were fed exclusively on ostracods throughout their developmental period. Before the beginning of the experiment with ostracods, seven *Triturus* larvae were randomly selected from the breeding aquarium and put in a tank with 1.2 L of fresh tap water for 48 h. During this period, they were not fed to ensure that water with predator chemical cues contained no alarm cues originating from consumed ostracods.

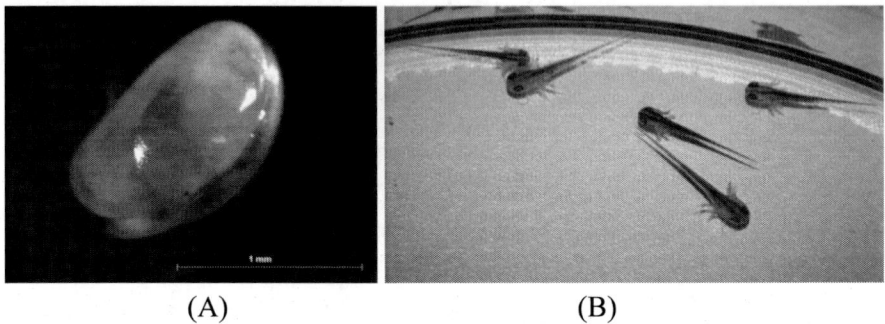

(A)   (B)

Figure 1. *Heterocypris incongruens* (A) and *Triturus* larvae (B). Images were obtained using a Zeiss Stemi DV4 stereomicroscope (Carl Zeiss, Munich, Germany) and AxioVision SE64 Rel. 4.9.1 software (ver. 4.2 for Windows).

"Water with alarm cues" was obtained by submerging and immediately crushing with a pestle 10 *H. incongruens* individuals in 5 mLl of water. The water was filtered through a "clip"-equipped folding grid with a solid border (grid size: 300 mesh × 83 µm pitch; hole width: 63 µm; open area size: 55%) after 5 min to ensure the water contained only alarm cues, but no tissue of dead ostracods.

Experimental Design

The experiments were carried out in a Petri dish measuring 9 cm in diameter divided by partial separators into eight open chambers radiating from the central part (4 cm in diameter).

The experimental arena was based on a design proposed by Schmit et al. (2007). The Petri dish was designed in a way to allow individuals free access to different items offered. Out of eight chambers in the Petri dish, different items were put in three of them: dark aquarium quartz sand in chamber 1, white aquarium quartz sand in chamber 3 and a cellulose sponge in chamber 6 (Figure 2). The rest of the chambers were left empty.

The quartz sand used had a grain size of 0.5 – 1 mm. It was intensely rinsed prior to the experiment. A cellulose sponge (5 mm in length, 8 mm in diameter) was used as a carrier of the test substances during the experiment. The dish was filled with water to a level such that it did not overflow the chambers' separators. In accordance with the called-for specifications*specifications*, the presented minerals were as follows: calcium, 80.0 mg/L; sodium, 6.5 mg/L; potassium, 1.0 mg/L; bicarbonates, 360.0 mg/L; sulphates, 12.6 mg/L; chlorides, 6.8 mg/L; nitrates, 3.7 mg/L; silicon, 15.0 mg/L; and magnesium, 26.0 mg/L.

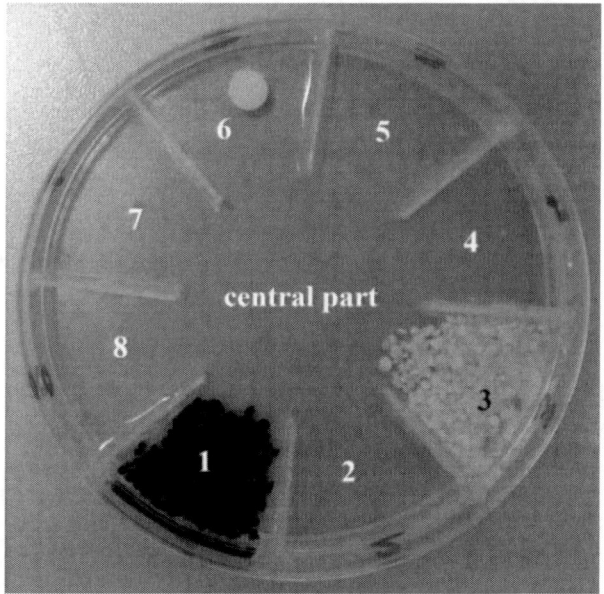

Figure 2. Experimental settings in the Petri dish filled with water: 1 – dark quartz sand, 3 – white quartz sand, 6 – cellulose sponge, 2, 4, 5, 7, 8 – empty chambers.

Two experimental groups and a control group were set up and scored. Behaviour of the ostracods was observed in response to: 1) the presence of assumed chemical compounds originating from the predator, designated as the "predator group"; and 2) alarm signals from injured conspecific ostracods, designated as the "alarm group". The experimental sponge was soaked with 0.5 mL of "water with predator chemical cues" or with "water with alarm cues".

Eight replicates were set up for each group, and 10 individuals were used for each replicate (240 individuals in all, 80 individuals per group). Ten individuals were simultaneously placed with a pipette in the central part of the Petri dish in order to have equal access to any of the chambers. The sponge in chamber 6 was placed at once with specimens in the central part of the Petri dish.

The experiments were performed at the same time of day (between 1 p.m. and 3 p.m.), ensuring that daily fluctuations in the laboratory environment had no effect on the ostracods' behaviour. To avoid any possible influence from the surroundings, the orientation of the Petri dish was randomized. The total number of individuals in the central part and each chamber of the Petri dish was noted every minute during the 10-minute period of observation.

Statistical Analysis

In all replicates, the total number of individuals in the central part and chambers of the Petri dish, expressed in terms of percentage, was compared within and between the control and experimental groups using the Z-test (Zar, 1999).

Data on the number of individuals in the central part of the Petri dish and chambers with different items were noted every minute during the 10-minute period of observation and checked for normality using the Kolmogorov-Smirnov test. Non-parametric Kruskal-Wallis ANOVA followed by the post-hoc Mann-Whitney U test was applied in order to analyse the number of individuals in the control and experimental groups. In addition, Friedman ANOVA followed by the Wilcoxon matched pairs test was used to analyse the number of individuals within each group. STATISTICA®, ver. 5.0 (StatSoft) was used for all statistical analyses.

RESULTS

Effects of Predator-Derived Chemical Cues and Conspecific Alarm Cues on Behaviour of *H. iIncongruens*

Different distribution of individuals between the central part and chambers of the Petri dish was observed within the control, predator and alarm groups, as well as between these groups (Figure 3).

After being placed in the central part of the Petri dish, a significant number of individuals of the control group (84.2%) moved into some of the chambers ($z = -38.75$, $p < 0.01$). The same pattern was observed in the alarm group ($z = -23.90$, $p < 0.01$), where 71.1% of individuals moved toward the chambers. Contrary to these groups, the behaviour of individuals in the predator group was different. To be specific, they were almost equally distributed in the central part and in chambers of the Petri dish (50.6% and 49.4%, respectively). Comparing the number of individuals recorded in the central part of the Petri dish, we see that this number was significantly higher in the predator and alarm groups than in the control group ($z = -20.95$, $p < 0.01$ and $z = -8.92$, $p < 0.01$, respectively). Also, the number of individuals recorded in the central part of the Petri dish was significantly higher in the predator group than in the alarm group ($z = 12.57$, $p < 0.01$).

Regarding the preference for some of the chambers supplied with different items, certain peculiarities within the control, predator and alarm groups, as well as between these groups, were observed (Figure 3). The largest number of individuals from the control group was recorded in the chamber with a cellulose sponge (about 25.6%). Individuals of the predator group were also most numerous in the chamber with a cellulose sponge (9.5%). Behaviour of individuals in the alarm group was quite different: they were largely distributed in the chamber with white quartz sand (19.0%). In all groups, the smallest number of individuals was recorded in the chamber filled with black quartz sand (0.5% in the control group, 0.8% in the predator group and 1.4% in the prey group) (Figure 3).

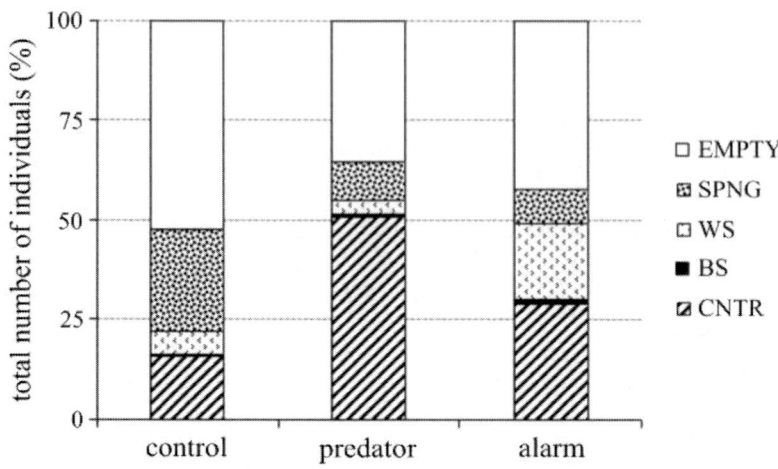

Figure 3. Total number of *H. incongruens* individuals (expressed in %) in the central part (CNTR) and in chambers with black quartz sand (BS), white quartz sand (WS) and a cellulose sponge (SPNG), as well as in empty chambers (EMPTY) of the Petri dish in the control, predator and alarm groups.

In addition, grouping of individuals in the form of a chain during the experiment was observed in the alarm group (Figure 4).

Table 1. Values of Z-test for within-group comparisons

Group	Item	BS	WS
Control	WS	- 8.53**	
	SPNG	- 21.09**	- 15.46**
Predator	WS	- 5.39**	
	SPNG	- 11.22**	- 6.88**
Alarm	WS	- 16.48**	
	SPNG	- 9.09**	8.86**

**p < 0.01; BS – black quartz sand, WS – white quartz sand, SPNG – cellulose sponge.

Differences in the number of individuals in the chambers with different items within each group and between the control and experimental groups are presented in Tables 1 and 2, respectively.

Table 2. Values of Z-test for between-group comparisons

Item	Group	Control	Predator
BS	Predator	- 0.90	
	Alarm	- 2.57**	- 1.72
WS	Predator	- 5.39**	
	Alarm	- 17.30**	- 13.87**
SPNG	Predator	11.99**	
	Alarm	13.10**	1.24

**$p < 0.01$, *$p < 0.05$; BS – black quartz sand, WS – white quartz sand, SPNG – cellulose sponge.

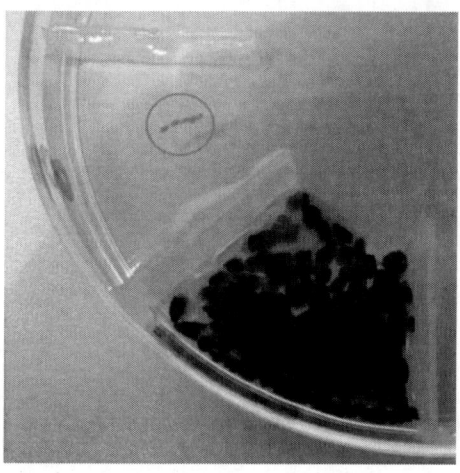

Figure 4. Formation of an aggregated chain in *H. incongruens* (in circle).

Time Course of Behavioural Changes Induced by Predator-Derived Chemical Cues and Conspecific Alarm Cues in *H. iIncongruens*

Some peculiarities of the behaviour of *H. incongruens* were revealed in the control and experimental groups during 10-minute period of observation (Table 3, Figure 5).

In the central part of the Petri dish, the number of individuals in the predator group was significantly higher during the whole period of observation. On the other hand, in the alarm group, this number was significantly higher only during the first three minutes of observation (both comparisons were made relative to the control group) (Figure 5, Table 4). The number of individuals of the predator group which occupied the centre of the Petri dish was also significantly higher in comparison with the alarm group from the fifth minute of observation until the end of observation.

In analysis of the distribution of individuals over time within each group, a significant difference was recorded only in the alarm group (Table 5). In this group, the largest number of individuals was observed in the centre in the first minute, after which it gradually decreased over the next five minutes to a level that was maintained until the end of observation (Table 6).

In the chamber filled with black quartz sand, the number of individuals in the control and both experimental groups was very small during the whole 10-minute period of observation (Figure 5), and statistical analysis revealed no significant differences between these groups (Table 3). A significant difference was recorded only within the predator group (Table 5), with an increased number of individuals in the last minute of observation.

Table 3. Results of Kruskal-Wallis ANOVA over time.

	CNTR	BS	WS	SPNG
	$H_{(2, 24)}$	$H_{(2, 24)}$	$H_{(2, 24)}$	$H_{(2, 24)}$
1'	11.91**	1.05	1.67	13.93***
2'	14.38***	2.00	3.70	13.05***
3'	14.07***	2.00	5.20	13.46***
4'	12.04**	0.00	14.05***	3.01
5'	9.03*	2.00	5.20	5.90
6'	10.24**	2.00	11.64**	3.17
7'	8.59*	2.19	9.83**	4.56
8'	7.31*	2.00	14.48***	9.14**
9'	6.65*	2.19	12.15**	7.49*
10'	11.53**	1.65	13.00**	5.73

*** $p < 0.001$, ** $p < 0.01$, * $p < 0.05$; CNTR – central part of the Petri dish, BS – black quartz sand, WS – white quartz sand, SPNG – cellulose sponge; H(degree of freedom, number of replicates).

In the chamber with white quartz sand, the number of individuals in both the predator group and the control group was the same during the first five minutes of observation and remained relatively low in both groups. On the contrary, the number of individuals in the alarm group was higher than in both the control group and the predator group during the entire 10-minute period of observation. A significant difference was observed in the fourth minute, and from the sixth minute to the end of observation (Table 4). This number gradually increased, reaching a maximum in the eighth minute of observation (Figure 5, Tables 5 and 6).

Table 4. Results of the Mann-Whitney U test over time

	Predator *vs.* Control			Alarm *vs.* Control		
	CNTR	WS	SPNG	CNTR	WS	SPNG
1′	5.0**	25.5	3.5**	3.5**	21.5	2.0***
2′	1.0***	31.5	5.0**	3.5**	18.5	4.0**
3′	0.0***	29.5	3.0**	6.0**	15.5	3.0**
4′	0.5***	32.0	20.5	16.5	5.5**	17.5
5′	5.5**	29.5	12.0*	23.0	15.0	14.0
6′	4.5**	19.5	19.0	27.5	10.5*	17.5
7′	8.0**	28.5	18.0	28.0	8.5**	14.0*
8′	11.0*	18.0	11.0*	24.0	6.5**	7.5**
9′	12.0*	18.5	11.0*	24.0	7.5**	9.5*
10′	3.0**	15.0*	12.0*	31.0	11.0*	15.5
Predator *vs.* Alarm						
	CNTR		WS		SPNG	
1′	30.0		26.5		30.0***	
2′	21.0		16.5		22.5***	
3′	19.5		15.0		29.0***	
4′	14.0		5.5**		26.0	
5′	12.0*		15.5		30.5	
6′	8.5*		5.0**		28.0	
7′	9.0*		7.0**		26.0	
8′	10.5*		1.0***		23.5**	
9′	11.5*		5.0**		31.5*	
10′	6.5**		3.0***		26.5	

*** $p < 0.001$, ** $p < 0.01$, * $p < 0.05$; CNTR – central part of the Petri dish, BS – black quartz sand, WS – white quartz sand, SPNG – cellulose sponge.

In the chamber with a cellulose sponge, the pattern of behaviour of individuals in the predator group and the alarm group was very similar (Figure 5). They did not prefer this chamber, but they did not completely avoid it.

Compared to the control group, reduction in the number of individuals was significant during the first three minutes in both groups; in the fifth minute and from the eighth minute until the end of observation in the predator group; and from the seventh minute to the ninth minute of observation in the alarm group (Table 4). In analysis of the distribution of individuals over time within each group, no significant differences were noticed (Table 5).

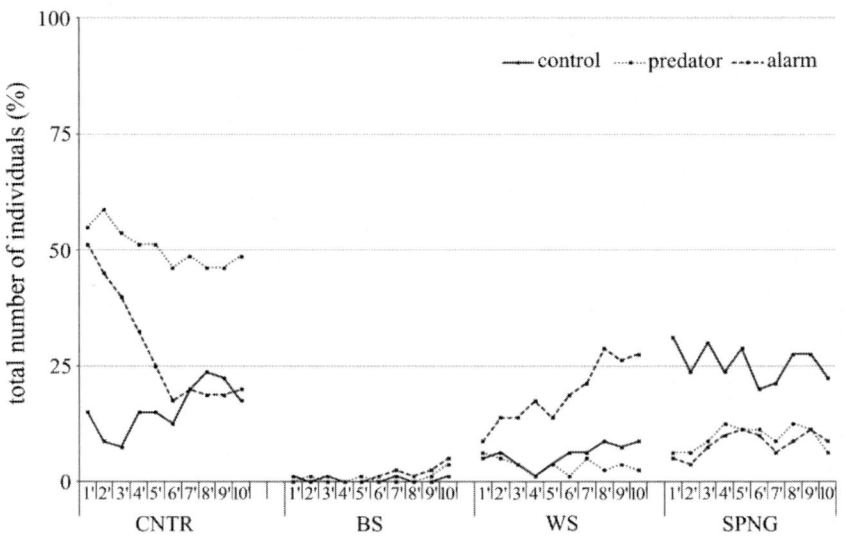

Figure 5. Number of *H. incongruens* individuals (expressed in %) during the 10-minute period of observation in the central part of the Petri dish (CNTR) and chambers with black quartz sand (BS), white quartz sand (WS) and a cellulose sponge (SPNG) in the control, predator and alarm groups.

Table 5. Results of Friedman ANOVA

	Control $\chi^2_{(9, 8)}$	Predator $\chi^2_{(9, 8)}$	Alarm $\chi^2_{(9, 8)}$
CNTR	11.99	11.32	28.29***
BS	6.35	18.00*	14.66
WS	12.49	9.34	27.62***
SPNG	8.50	8.70	8.79

*** $p < 0.001$, * $p < 0.05$; CNTR – central part of the Petri dish, BS – black quartz sand, WS – white quartz sand, SPNG – cellulose sponge; χ^2(degree of freedom, number of replicates).

Table 6. Results of Wilcoxon matched pairs test of the alarm group over time

	CNTR				WS				
	1′	2′	3′	4′	1′	2′	3′	4′	5′
2′	- 1.7				- 1.6				
3′	- 1.8	- 1.4			- 1.1	0.0			
4′	- 2.1*	- 1.9	- 1.9		- 2.1*	- 1.0	- 1.7		
5′	- 2.1*	- 2.2*	- 1.9	- 1.3	- 2.0*	0.0	0.0	- 1.1	
6′	- 2.3*	- 2.4*	- 2.0*	- 2.1*	- 2.3*	- 1.4	- 1.6	- 0.5	- 1.6
7′	- 2.3*	- 2.2*	- 2.0*	- 1.7	- 2.6**	- 2.1*	- 1.4	- 1.0	- 2.1*
8′	- 2.2*	- 2.3*	- 2.0*	- 1.6	- 2.4*	- 2.2*	- 2.4*	- 2.3*	- 2.1*
9′	- 2.1*	- 2.1*	- 1.9	- 1.4	- 2.3*	- 2.2*	- 1.8	- 1.4	- 1.6
10′	- 2.2*	- 2.1*	- 1.9	- 1.4	- 2.1*	- 1.8	- 1.7	- 1.3	- 1.6

**$p < 0.01$, *$p < 0.05$; CNTR – central part of the Petri dish, WS – white quartz sand.

DISCUSSION

During their long evolutionary history, ostracods developed various antipredator strategies that contributed to the success, diversity and wide cosmopolitan distribution of this group of crustaceans (Uiblein et al., 1994; Aarnio and Mattila, 2000; Kiss, 2004; Matzke-Karasz et al., 2014). They perceive infochemicals and use them as a cue for spatial orientation, various interactions, determining the presence of food and locating it (Rittschof, 1992; Rossi et al., 2011; Miličić et al., 2015). Often, their behavioural response represents a trade-off between antipredator and foraging activities. In the presence of a predator, the potential prey uses a wide range of strategies and activities conferring the ability to avoid or offset attack, such as changes in morphology and/or physiology, modifications of post-hatching behaviour or alteration of life history traits (Uiblein et al., 1994).

Results of the present study indicate changes in the behaviour of *H. incongruens* due to the assumed presence of various chemical compounds (kairomones from predators or alerting substances of injured conspecifics). Perception of chemical cues induced a specific form of defensive behaviour. Specimens exposed to predator-derived cues stayed longer in the centre of the Petri dish, remaining immobile ["playing dead" behaviour, according to Schmit et al. (2007)].

A lower level of activity can ensure higher survival rates by making prey less conspicuous to predators (Skelly, 1994). Vandekerkhove et al. (2012) recorded migration of *H. incongruens* to open water in the presence of predators, accompanied by reduction of swimming activities, which agrees with the results obtained in our study.

We found that *H. incongruens* responded in a similar manner in both experimental groups, which indicates that both types of chemical cues (predator-derived molecules and alarming substances originating from injured conspecifics) warned other individuals of the existence of immediate danger. In both treatments, the chamber with the source of alarming substances was not completely avoided by the ostracods, but it was significantly less frequently visited than in the control group. Although predator avoidance and reduced locomotion appear to be universal antipredator mechanisms in ostracods (Vandekerkhove et al., 2012), the short-term trials conducted in the present study showed that individuals of *H. incongruens* need some time to evaluate environmental cues. They are able to modify behaviour and defensive strategy in a short period of time, the length of which depends on the infochemicals perceived.

Grouping of individuals and aggregating in the form of a chain was observed when experimental specimens were exposed to the scents of injured conspecific individuals (a sign of high hazard). It is possible that formation of such aggregations represents part of a defence strategy with the aim of creating a "massive body", which then may be less attractive to potential predators. To be specific, a larger body of individuals is known to be a limiting factor in consumption of a number of crustaceans as prey. In *Daphnia*, for example, increasing body size and various morphological structures can significantly increase the survival rate. This represents an inducible antipredator defence that evolved in coexistence with their ancient predators (Petrusek et al., 2009; Rabus and Laforsch, 2011). Also, aggregation and reduced activity can increase the chances of such a formation being considered as a sort of a non-living object.

In the presence of conspecific alarm cues, ostracods preferred the light-coloured sand, probably because they perceive this substrate as a potential hiding ground. However, such behaviour was not recorded when the specimens were exposed to predator-derived cues, nor in the control group. All of the considerations mentioned above lead us to speculate that the releasing of body fluids of injured animals into the surrounding medium represents a much more potent signal of danger for *H. incongruens*, triggering a more pronounced defensive reaction than the smell of an invisible predator itself. It seems that harm avoidance constitutes a significant part of the defensive strategy of *H. incongruens*. From that point of view, it should be noted that the importance of camouflage has been stressed by many authors (Alcock, 2005; Stevens and Merilaita, 2011; Stevens et al., 2011; Dugatkin, 2013; Troscianko et al., 2013; Nordell and Valone, 2015). Some species of ostracods flee, hide among underwater plants or search for a protective substrate, thereby decreasing the predation rate (Roca et al., 1993; Uiblein et al., 1994; Kiss, 2004). *H.eterocypris incongruens* lives in shallow waters with good illumination, where sunlight commonly reaches the very bottom of the habitat. This may explain why individuals prefer lighter-coloured ground (chambers with white quartz sand in our experiment), where they are less visible to predators, using it as a shelter when they are in danger. It could also be the reason why *H. incongruens* strongly avoided black quartz sand in all experimental trials. Moreover, it is is possible that individuals perceive a dark background (and avoid it) as a kind of "physical barrier" or simply as part of a habitat "unsuitable for life". Further research is needed to determine the true meaning of such behaviour in selection of habitats in freshwater environments, since the background colour might have high ethological and ecological significance in some species of this crustacean group (Nakamura, 1954).

The existence of a wide spectrum of different strategies in the presence of predators has high importance for survival. Such responses are different for different predators, and can vary widely among populations (Templeton and Shriner, 2004; Lakowitz et al., 2008). It seems that a mix of various types of alarm substances is necessary for inducing a complete antipredator response (Schoeppner and Relyea, 2005; Schoeppner and Relyea, 2009). Poor knowledge of the chemical structure of compounds that lead to such responses, their extremely low concentrations and the difficulty of isolating kairomones from natural aquatic habitats further complicate the goal of reaching correct conclusions (Mitchell et al., 2017). Despite recent discoveries in the field of behavioural ecology and chemoecology, we are just beginning to understand the mechanisms governing the complex forms of antipredator behaviour in aquatic organisms.

CONCLUSION

Our findings support the idea that chemosensory alarm cues induce an antipredator behavioural response in ostracods, even when the predator is not physically/visually present. We assume that individuals of *H. incongruens* recognized the existence of even an invisible predator because the predator released its own odours into the experimental medium. The odours could also be chemical compounds and by-products of digestion. The ~~antipredator~~ behavioural response was even more pronounced in the presence of conspecific chemical alarm cues. In other words, body contents leaked from injured ostracods triggered more intense modification of behaviour, thereby stressing the importance of chemical alarm cues for the realization of a plastic response in crustaceans (James and McClintock, 2017). It would be particularly interesting to further investigate the incidence of grouping behaviour in the presence of conspecific alarm cues, especially combined with predator-derived infochemicals. This could be a risky undertaking in natural conditions. Our study confirmed that *H. incongruens* can be a useful model system, since it is easily raised in laboratory conditions, and the small size of individuals making experiments relatively easy to conduct.

ACKNOWLEDGMENTS

This work was supported by the Ministry of Education, Science and Technological Development of the Republic of Serbia (Grant No. 176019). The authors are very grateful to Mr. Raymond Dooley for proofreading the text and improving the English. We also thank Tijana Vučić (Department of Morphology, Systematics and Phylogeny of Animals, Faculty of Biology, University of Belgrade) for obtaining the newt larvae used in the experiment, and to Dr. Tamara Karan Žnidaršič and Dr. Sara Selaković for reading a draft version of the manuscript. The rearing of *Triturus* spp. sp. for the experiment was approved by the Ethical Committee of the Institute for Biological Research "Siniša Stanković" Institute for Biological Research, University of Belgrade (document No. 03 – 03/16). All animals were collected in accordance with a permit (No. 353 01 2014 08) provided by the Ministry of Energy, Development and Environmental Protection of the Republic of Serbia.

REFERENCES

Aarnio, K. & Mattila, J. (2000). Predation by juvenile *Platichthys flesus* (L.) on shelled prey species in a bare sand and a drift algae habitat. *Hydrobiologia, 440* (1), 347-355.

Alcock, J. (2005). *Animal Behavior.* Sunderland: Sinauer Associates Inc.

Barreto, R. E., Miyai, C. A., Sanches, F. H., Giaquinto, P. C., Delicio, H. C. & Volpato, G. L. (2013). Blood cues induce antipredator behavior in *Nile tilapia* Nile tilapia conspecifics. *PLoS One, 8* (1), e54642.

Delorme, L. D. (1991). Ostracoda. In H. Thorp and A. P. Covich (Eds.), *Ecology and Classification of North American Freshwater Invertebrates*, (pp. 691-722). San Diego: Academic Press.

Dugatkin, L. A. (2013). *Principles of Animal Behavior.* New York: W. W. Norton & Company, Inc.

Edmunds, M. (1974). *Defence in Animals: a Survey of Anti-predator Defences.* Harlow: Longman.

Endler, J. A. (1995). Multiple-trait coevolution and environmental gradients in guppies. *Trends in Ecology & Evolution, 10* (1), 22-29.

Fasola, M. & Canova, L. (1992). Feeding habits of *Triturus vulgaris, T. cristatus* and *T. alpestris* (Amphibia, Urodela) in the northern Apennines (Italy). *Bollettino di Zoologia, 59* (3), 273-280.

Fernandez, R., Nandini, S., Nandini, S. S. S. & Castellanos-Páez, M. E. (2016). Demographic responses of *Heterocypris incongruens* (Ostracoda) related to stress factors of competition, predation and food. *Journal of Limnology, 75* (1s), 31-38.

Ferrari, M. C. O., Wisenden, B. D. & Chivers, D. P. (2010). Chemical ecology of predator-prey interactions in aquatic ecosystems: a review and prospectus. *Canadian Journal of Zoology, 88* (7), 698-724.

Fryer, G. (1997). The horse-trough ostracod *Heterocypris incongruens*. *Naturalist, 122,* 121-135.

Ganning, B. (1971). On the ecology of *Heterocypris salinus, H. incongruens* and *Cypridopsis aculeata* (Crustacea: Ostracoda) from Baltic brackish-water rockpools. *Marine Biology, 8* (4), 271-279.

Harvey, T. H. P., Velez, M. I. & Butterfield, N. J. (2012). Exceptionally preserved crustaceans from western Canada reveal a cryptic Cambrian radiation. *PNAS, 109* (5), 1589-1594.

Havel, J. E., Link, J. & Niedzwiecki, J. (1993). Selective predation by *Lestes* (Odoiiata, Lestidae) on littoral microcrustacea. *Freshwater Biology, 29* (1), 47-58.

Hazlett, B. A. (2011). Chemical cues and reducing the risk of predation. In T. Breithaupt and M. Thiel (Eds.), *Chemical Communication in Crustaceans,* (pp. 355-370). New York: Springer.

Holmes, J. A. & Chivas, A. R. (2002). *The Ostracoda: Applications in Quaternary Research.* Geophysical Monograph Series, 131. Washington, DC: American Geophysical Union.

Horne, F. R. (1993). Survival strategy to escape desiccation in a freshwater ostracod. *Crustaceana, 65* (1), 53-61.

James, W. R. & McClintock, J. B. (2017). Anti-predator responses of amphipods are more effective in the presence of conspecific chemical cues. *Hydrobiologia, 797* (1), 277-288.

Johnson, J. H. (1995). Diel feeding ecology of three species of aquatic insects. *Journal of Freshwater Ecology, 10* (2), 183-188.

Kiss, A. (2004). Field and laboratory observations on the microhabitat and food selection as well as predator avoidance of *Notodromas monacha* (Crustacea: Ostracoda). *Revista Española de Micropaleontología, 36* (1), 147-156. [*Spanish Journal of Micropaleontology*]

Külköylüoğlu, O., Dügel, M., Balci, M., Deveci, A., Avuka, D. & Kiliç, M. (2010). Limnoecological relationships between water level fluctuations and Ostracoda (Crustacea) species composition in Lake Sünnet (Bolu, Turkey). *Turkish Journal of Zoology, 34* (4), 429-442.

Lakowitz, T., Brönmark, C. & Nyström, P. (2008). Tuning in to multiple predators: conflicting demands for shell morphology in a freshwater snail. *Freshwater Biology, 53* (11), 2184-2191.

Lass, S. & Spaak, P. (2003). Chemically induced anti-predator defences in plankton: a review. *Hydrobiologia, 491* (1-3), 221-239.

Liperovskaya, E. S. (1948). On the feeding of freshwater ostracods. *Zoologicheskie Zhurnal, 27* (2), 125-136.

Matzke-Karasz, R., Nagler, C. & Hofmann, S. (2014). The ostracod springtail – camera recordings of a previously undescribed high-speed escape jump in the genus *Tanycypris* (Ostracoda, Cypridoidea). *Crustaceana, 87* (8-9), 1072-1094.

Mesquita-Joanes, F., Smith, A. J. & Viehberg, F. A. (2012). The ecology of Ostracoda across levels of biological organisation from individual to ecosystem: a review of recent developments and future potential. In D. J. Horne, J. A. Holmes, J. Rodríguez-Lázaro and F. Viehberg (Eds.), *Ostracoda as Proxies for Quaternary Climate Change*, Chapter 2. Amsterdam: Elsevier. *Developments in Quaternary Science, 17*, 15-35.

Miličić, D. M., Majstorović, A. P., Pavković-Lučić, S. B. & Savić, T. T. (2015). Behaviour and food selection of *Heterocypris incongruens* (Ostracoda). *Crustaceana, 88* (10-11), 1097-1110.

Mitchell, M. D., Bairos-Novak, K. R. & Ferrari, M. C. O. (2017). Mechanisms underlying the control of responses to predator odours in aquatic prey. *Journal of Experimental Biology*, *220* (Pt 11), 1937-1946.

Moguilevsky, A. & Gooday, A. J. (1977). Some observations on the vertical distribution and stomach contents of *Gigantocypris muelleri* Skogsberg 1920 (Ostracoda, Myodocopina). In H. Löffler and D. Danielopol (Eds.), *Aspects of Ecology and Zoogeography of Recent and Fossil Ostracoda*, (pp. 263-270). The Hague: W. Junk.

Nakamura, N. (1954). Study on the ecology of *Cypridina hilgendorfii*. In Japanese Society of Fisheries (Ed.), *General View of Fisheries*, (pp. 108-127). Tokyo: Japanese Association for the Advancement of Science.

Nordell, S. E. & Valone, T. J. (2015). *Animal Behavior: Concepts, Methods, and Applications.* New York: Oxford University Press.

Orr, M. V., Hittel, K. & Lukowiak, K. (2009). 'Different strokes for different folks': geographically isolated strains of *Lymnaea stagnalis* only respond to sympatric predators and have different memory forming capabilities. *Journal of Experimental Biology*, *212* (Pt 14), 2237-2247.

Ottonello, D. & Romano, A. (2011). Ostracoda and Amphibia in temporary ponds: who is the prey? Unexpected trophic relation in a Mediterranean freshwater habitat. *Aquatic Ecology*, *45* (1), 55-62.

Petrusek, A., Tollrian, R., Schwenk, K., Haas, A. & Laforsch, C. (2009). A "crown of thorns" is an inducible defense that protects *Daphnia* against an ancient predator. *Proceedings of the National Academy of Sciences*, *106* (7), 2248-2252.

Rabus, M. & Laforsch, C. (2011). Growing large and bulky in the presence of the enemy: *Daphnia magna* gradually switches the mode of inducible morphological defences. *Functional Ecology*, *25* (5), 1137-1143.

Reichholf, J. (1983). Okologie und verhalten des muschelkrebses *Heterocypris incongruens* Claus, 1892. *Spixiana*, *6*, 205-210. [Ecology and behavior of seed shrimps *Heterocypris incongruens* Claus, 1892. *Spixiana*, *6*, 205-210].

Rittschof, D. (1992). Chemosensation in the daily life of crabs. *American Zoologist*, *32* (3), 363-369.

Roca, J. R., Baltanas, A. & Uiblein, F. (1993). Adaptive responses in *Cypridopsis vidua* (Crustacea: Ostracoda) to food and shelter offered by a macrophyte (*Chara fragilis*). *Hydrobiologia*, *262* (2), 127-131.

Rossi, V., Benassi, G., Belletti, F. & Menozzi, P. (2011). Colonization, population dynamics, predatory behaviour and cannibalism in *Heterocypris incongruens* (Crustacea: Ostracoda). *Journal of Limnology*, *70* (1), 102-108.

Rossi, V., Martorella, A. & Menozzi, P. (2013). Hatching phenology and voltinism of *Heterocypris barbara* (Crustacea: Ostracoda) from Lampedusa (Sicily, Italy). *Journal of Limnology*, *72* (2), 227-237.

Schmit, O., Rossetti, G., Vandekerkhove, J. & Mezquita, F. (2007). Food selection in *Eucypris virens* (Crustacea: Ostracoda) under experimental conditions. *Hydrobiologia*, *585* (1), 135-140.

Schoeppner, N. M. & Relyea, R. A. (2005). Damage, digestion, and defence: the roles of alarm cues and kairomones for inducing prey defences. *Ecology Letters*, *8* (5), 505-512.

Schoeppner, N. M. & Relyea, R. A. (2009). Interpreting the smells of predation: how alarm cues and kairomones induce different prey defences. *Functional Ecology*, *23* (6), 1114-1121.

Skelly, D. K. (1994). Activity level and the susceptibility of anuran larvae to predation. *Animal Behaviour*, *47* (2), 465-468.

Stevens, M. & Merilaita, S. (2011). *Animal Camouflage: Mechanisms and Function.* Cambridge: Cambridge University Press.

Stevens, M., Searle, W. T. L., Seymour, J. E., Marshall, K. L. & Ruxton, G. D. (2011). Motion dazzle and camouflage as distinct anti-predator defenses. *BMC Biology*, *9* (1), 81.

Templeton, C. N. & Shriner, W. M. (2004). Multiple selection pressures influence Trinidadian guppy (*Poecilia reticulata*) antipredator behavior. *Behavioral Ecology*, *15* (4), 673-678.

Tollrian, R. & Harvell, C. D. (1999). *The Ecology and Evolution of Inducible Defenses.* Princeton, NJ: Princeton University Press.

Troscianko, J., Lown, A. E., Hughes, A. E. & Stevens, M. (2013). Defeating crypsis: detection and learning of camouflage strategies. *PLoS One, 8* (9), e73733.

Uiblein, F., Roca, J. R. & Danielopol, D. L. (1994). Experimental observations on the behaviour of the ostracode *Cypridopsis vidua*. *Verhandlungen des Internationalen Verein Limnologie, 25* (4), 2418-2420.

Vandekerkhove, J., Martens, K., Rossetti, G., Mesquita-Joanes, F. & Namiotko, T. (2013). Extreme tolerance to environmental stress of sexual and parthenogenetic resting eggs of *Eucypris virens* (Crustacea, Ostracoda). *Freshwater Biology, 58* (2), 237-247.

Vandekerkhove, J., Namiotko, T., Hallmann, E. & Martens, K. (2012). Predation by macroinvertebrates on *Heterocypris incongruens* (Ostracoda) in temporary ponds: impacts and responses. *Fundamental and Applied Limnology, 181* (1), 39-47.

Vannier, J., Abe, K. & Ikuta, K. (1998). Feeding in myodocopid ostracods: functional morphology and laboratory observations from videos. *Marine Biology, 132* (3), 391-408.

Wilkinson, I. P., Wilby, P. R., Williams, M., Siveter, D. J. & Vannier, J. (2007). Ostracod carnivory through time. In A. M. T. Elewa (Ed.), *Predation in Organisms: A Distinct Phenomenon*, (pp. 39-57). Berlin, Heidelberg: Springer.

Williams, M., Siveter, D. J., Salas, M. J., Vannier, J., Popov, L. E. & Pour, M. G. (2008). The earliest ostracods: the geological evidence. *Senckenbergiana lethaea, 88* (1), 11-21.

Wilson, R. S., Kraft, P. G. & Van Damme, R. (2005). Predator-specific changes in the morphology and swimming performance of larval *Rana lessonae*. *Functional Ecology, 19* (2), 238-244.

Zar, J. H. (1999). *Biostatistical Analysis.* New Delhi: Pearson Education.

In: The Zoological Guide to Crustacea ISBN: 978-1-53616-366-7
Editor: Noelle Lachance © 2019 Nova Science Publishers, Inc.

Chapter 4

A ZOOLOGICAL OVERVIEW OF THE LARVAE OF DIFFERENT GROUPS OF LOBSTERS

Brady K. Quinn[*]
Department of Biological Sciences,
University of New Brunswick, Saint John, NB, Canada

ABSTRACT

Marine decapod crustaceans that crawl ('Reptantia'), brood their eggs (Pleocyemata), and are not considered 'crabs' or 'shrimps' are generally termed 'lobsters'. These include members of 6-7 different decapod infraorders, including the clawed nephropoid and reef lobsters (Astacidea: Nephropoidea and Enoplometopoidea, respectively), blind lobsters (Polychelida), ghost or mud lobsters ('Thalassinidea' = Axiidea and Gebiidea), clawless spiny, slipper, and furry lobsters (Achelata: Palinuridae, Scyllaridae, and 'Synaxidae,' respectively), and the 'living fossil' glypheid lobsters (Glypheidea), as well as certain crabs that are also commonly termed 'squat lobsters' (certain members of Anomura: Chirostyloidea and Galatheoidea). Many lobsters support important fisheries, and they play important ecological roles as large, benthic consumers in marine communities. All lobsters develop through a pelagic

---

[*] Corresponding Author's E-mail: bk.quinn@unb.ca.

(planktonic and/or pelagic) larval phase, which differs considerably from the benthic juvenile and adult phases of their life cycles in terms of its morphology, behavior, physiology, and ecology. The larval phase is vital to the maintenance of lobster populations and fished lobster stocks as it supplies new recruits to benthic populations, but for some taxa relatively little is known about their larval development. This chapter briefly reviews the larval phases of all major lobster taxa, with particular emphasis on the variations in larval developmental patterns, morphology, life cycle characteristics (duration, type and number of phases and stages, etc.), and behaviors while in the water column across taxa. Lobsters develop through two or three larval phases after hatching that are comprised of one to several different stages each. A short-lived 'pre-zoea' stage has been observed after hatching in nearly all lobster taxa, but its role and status within the life cycle remains unclear. This is followed by varying numbers of zoea (mysis) stages after hatching, which in some cases are highly modified, as in the phyllosoma larvae of the Achelata and the 'eryoneicus' larvae of the Polychelida. The zoeal stages are then followed by a decapodid or 'postlarva' stage, which is a strong swimmer in some taxa and eventually settles to the benthos. This chapter demonstrates the diversity of larval phases and their importance to lobster populations and fisheries, and provides a summary of larval developmental patterns and behaviors and their potential impacts on larval dispersal and lobster demographic connectivity.

Keywords: Achelata, Astacidea, Axiidea, Chirostyloidea, Galatheoidea, Gebiidea, lobster, larva, Polychelida, vertical migratory behavior

1. INTRODUCTION

A diverse array of decapod crustaceans in the suborder Pleocyemata have historically been known by the colloquial term 'lobster'. In its most traditional and original sense, the word was applied to the clawed homarid (*Homarus* spp.) lobsters, in the family Nephropidae in the infraorder Astacidea. Its usage was later extended to apply to almost any crawling ('Reptantia,' as opposed to 'Natantia' (swimming 'shrimp')) marine decapod with claws (well-developed chelipeds) not demonstrating the 'carcinized' morphology (i.e., broadened cephalothorax and reduced abdomen) of a 'crab,' and even came to include clawless decapods (spiny and slipper lobsters) as well (Phillips et al. 1980; Quinn 2019; Wolfe et al. 2019). Depending on the source consulted, reptant marine decapods belonging to 6-7 different infraorders may be considered 'lobsters' (Holthuis 1991; Williams 1986, 1988; Chan 2010; Bracken-Grissom et al. 2014; Martin 2014a, b; Figure 1A-L). These groups are briefly outlined below, and the present chapter focused on their larvae.

The infraorder Astacidea contains the 'true' clawed lobsters (superfamily Nephropoidea: family Nephropidae, including the homarid species and their relatives; Figure 1A), reef lobsters (superfamily Enoplometopoidea: family Enoplometopidae; Figure 1B), and the freshwater crayfishes (superfamilies Astacoidea and Parastacoidea; Figure 1C) (Gherardi et al. 2010; Wahle et al. 2012; Goy 2014a). In all members of this group, the first (most anterior) pair of walking legs (pereiopods) is modified into a pair of well-developed, strong chelae or claws (chelipeds), and the second pair of walking legs is also chelate in the homarid lobsters (Wahle et al. 2012). While the homarid lobsters are quite large and inhabit a wide range of depths on mostly rocky substrates along the continental shelf (Lawton and Lavalli 1995; Wahle et al. 2013), other nephropoid species are smaller and inhabit deeper waters, often in soft-bottomed habitats (Bell et al. 2013; Johnson and Johnson 2013). The reef lobsters are also somewhat small decapods, but are brightly colored and inhabit shallow, well-lit coastal waters, often living on coral reefs (Wahle et al. 2012; Goy 2014a). Crayfishes superficially resemble nephropoid lobsters, but live associated with fresh water (with aquatic, amphibious, brackish, or terrestrial life habits), and thus are not considered lobsters (Williams 1988; Gherardi et al. 2010); they also completely lack free-living larvae (Goy 2014a), and thus are not discussed much herein.

The infraorder Achelata contains the clawless spiny or rock lobsters (family Palinuridae; Figure 1D) and slipper or shovel-nosed lobsters (family Scyllaridae; Figure 1E) (Phillips et al. 1980, 1994; Phillips and Kittaka 2000; Lavalli and Spanier 2010; Palero et al. 2014a), as well as the furry or coral lobsters (formerly given their own family, 'Synaxidae,' but now known to be nested within the Palinuridae; Davie 1990; Figure 1F). As the name implies, these lobsters completely lack chelae, but do have uniquely modified antennae that are either long and spiked at the base for defense (Palinuridae, expect furry lobsters in which they are reduced) or short and flattened into an airfoil-like structure used in swimming maneuvers (Scyllaridae) (Holthuis 1991; Lavalli and Spanier 2010). All these lobsters have extremely tough, durable carapaces, and share a uniquely modified type of larva, the phyllosoma, described in detail below (Phillips and Sastry 1980; Sekiguchi 1988; Baisre 1994; Palero et al. 2014a). Many inhabit coastal coral reefs, but others live in deep waters, and many are cave-dwelling and/or socially gregarious (Phillips et al. 1980, 1994; Holthuis 1991; Phillips and Kittaka 2000; Lavalli and Spanier 2010).

The infraorders Axiidea (Figure 1I) and Gebiidea (Figure 1J), which were formerly grouped together in the infraorder 'Thalassinidea', contain a wide array of species known as lobster shrimps, ghost shrimps, ghost lobsters, mud lobsters, etc. (Sakai 2004; Pohle et al. 2011; Dworschak et al. 2012; Pohle and Santana 2014). Most species have well-developed chelipeds and inhabit coastal areas, where they burrow in sediments or along the shores of embayments, and can be important bioturbators or sources of erosion (Dworschak et al. 2012; Saigusa et al. 2018). Thalassinids were not traditionally considered lobsters by some workers (e.g., Phillips et al. 1980; Williams 1986, 1988; Chan 2010; Bracken-Grisson et al. 2014), but have been treated as such by others (Holthuis 1991; Quinn 2019), so they are therefore discussed herein.

The infraorder Polychelida contains a unique group of eyeless, deep-sea lobsters, in which all or nearly all pereiopods are chelate in adults (Boyko 2006; Lavalli and Spanier 2010; Martin 2014c; Figure 1G). The 'eryoneicus' larvae of this group were previously thought to belong to different species from the adults due to their different and highly modified morphology (Martin 2014c).

The infraorder Glypheidea is a primitive group of lobster-like decapods, most of which are extinct, but with two 'living fossil' species (*Neoglyphea inopinata* and *Laurentaeglyphea neocaledonica*) known (Forest 2006; Wahle et al. 2012; Figure 1H). The larvae of glypheid lobsters have never been observed (Martin 2014b), so discussion of them in this chapter is limited.

Figure 1. Examples of adults of various lobster taxa: (A) *Homarus americanus* (Astacidea: Nephropoidea); (B) *Enoplometopus occidentalis* (Astacidea: Enoplometopoidea); (C) a freshwater crayfish, *Orconectes limosus* (Astacidea: Astacoidea); (D) *Panulirus interruptus* (Achelata: Palinuridae); (E) *Scyllarides aequinoctialis* (Achelata: Scyllaridae); (F) *Palinurellus wieneckii* (Achelata: Palinuridae (formerly 'Synaxidae')); (G) *Stereomastis sculpta* (Polychelida); (H) *Neoglyphea inopinata* (Glypheidea); (I) *Pestarella tyrrhena* (Axiidea (formerly 'Thalassinidea')); (J) *Upogebia major* (Gebiidea (formerly 'Thalassinidea')); (K) *Munidopsis tridens* (Anomura: Galatheoidea); (J) *Kiwa hirsuta* (Anomura: Chirostyloidea). Photo credits: (A, D, G, K) U.S. National Oceanic and Atmospheric Administration (NOAA), public domain; (B) D. Azuma (OpenCage), license CC BY-SA 2.5; (C) A.R. Thompsen, public domain; (E) B. Willis (Atsme), license CC BY-SA 4.0; (F) J. Poupin (Legall and Poupin 2019); (H) Citron, license CC BY-SA 3.0; (I) H. Hillewaert, license CC-BY-SA 4.0; (J) Anonymous, license CC-BY-SA 3.0; (L) Oregon State University (https://flickr.com/photos/33247428@N08/6439209127), license CC-BY-SA 2.0. All images obtained from Wikimedia Commons (https://commons.wikimedia.org/) except for (F).

Lastly, certain hermit crabs in the infraorder Anomura are commonly termed 'squat lobsters' (certain members of the superfamilies Chirostyloidea (Figure 1L) and Galatheoidea (Figure 1K)) (Williams 1988; Macpherson and Baba 2011; Tudge et al. 2012; Harvey et al. 2014). These crabs have a somewhat larger abdomen or 'tail' than that of true crabs and other hermit crabs, which is the main reason for their designation as 'lobsters' (Williams 1988). Species of squat lobsters occur in shallow coastal and deep-sea waters, including the yeti crabs (*Kiwa* spp.) that inhabit deep-sea hydrothermal vents (Lovrich and Thiel 2011; Macpherson and Baba 2011). Some species are also pelagic as adults, and form large swarms that swim throughout the water column (Haye et al. 2010; Wehrtmann and Acuña 2011).

Overall, lobsters are a familiar type of seafood that contributes to human nutrition throughout the world. Lobster fisheries can be extremely economically important, supporting large industries, personal livelihoods, and coastal communities and economies. Lobsters belonging to the infraorders Astacidea (Nephropoidea) and Achelata (various palinurids and scyllarids), as well as several squat lobsters (Anomura), support major fisheries around the globe. The largest and most lucrative of these are currently those targeting the nephropoids *Homarus americanus* and *H. gammarus* (Wahle et al. 2013) and *Nephrops norvegicus* (Bell et al. 2013), as well as spiny and rock lobsters in the genera *Panulirus* (Phillips et al. 2013a), *Palinurus* (Groeneveld et al. 2013), and *Jasus* (Jeffs et al. 2013), which generate many millions of dollars each annually. In some regions, such as northeastern North America, the lobster fishery has become the most valuable fishery, and the main or only fishery sustaining communities that traditionally relied on finfisheries (Wahle et al. 2013). Squat lobster fisheries thus far have only developed on a commercial scale in Latin America, mainly targeting species of *Pleuroncodes*, but given the high abundances of many squat lobsters there is much potential for the further development of fisheries targeting them (Wehrtmann and Acuña 2011). Smaller mixed-species or recreational fisheries, including those for slipper lobsters (Spanier and Lavalli 2007, 2013) and other nephropoid and palinurid species (Phillips et al. 1980, 2006), also exist and can be quite locally important, sometimes even contributing to the subsistence of coastal communities (Bell et al. 2013; Groeneveld et al. 2013; Jeffs et al. 2013). The importance of the above lobsters to humans is thus considerable. In addition, all lobsters are relatively large secondary consumers that act as important predators and scavengers in marine communities, and thus have importance to ecological communities (Phillips et al. 2013b). Indeed, in deep, oxygen-poor waters, squat lobster are frequently the first and sometimes the only scavengers that arrive to feed on animal carcasses, playing an important role in decomposition and nutrient cycling in the deep sea and between it and other marine realms (Lovrich and Thiel 2011). Burrowing activities by thalassinids and some nephropoid lobsters (e.g., *N. norvegicus*) can also impact the physical structure of benthic habitats (Bell et al. 2013; Johnson and Johnson 2013; Saigusa et al. 2018).

Almost all of the above lobsters have a biphasic life cycle that includes the familiar benthic juvenile and adult phases, the latter of which is targeted by fisheries, and a pelagic or planktonic larval phase (Phillips and Sastry 1980; Philips et al. 1980; Martin 2014b). Larvae represent new inputs of recruits into benthic populations of lobsters, and are thus an important component of the life cycles of these economically and ecologically important lobsters. The survival of larvae in a given year impacts the number of new recruits to a benthic population, which will impact the catches of the fishery a few years later (Ennis 1995; Lawton and Lavalli 1995; Phillips et al. 2006). Because they inhabit a fluid environment and are exposed to oceanographic currents, larvae are also capable of considerable potential dispersal. Excessive dispersal away from suitable settlement habitat has the potential to reduce larval survival to settlement and benthic recruitment (Incze et al. 2010; Pineda and Reyns 2018; Gendron et al. 2019). Dispersal of larvae may also connect different benthic populations of the same lobster species, as it creates the possibility that larvae hatched from one population can settle and recruit to another, different one (Philips and Sastry 1980; Ennis 1995; Phillips et al. 2006; Sekiguchi et al. 2007; Segura-Garciá et al. 2019). This can result in source-sink dynamics among populations or sub-populations, metapopulation-like structure, and difficulties for fisheries managers trying to predict future fisheries trends based on local information only (Chassé and Miller 2010; Kough et al. 2013; Quinn et al. 2017; Singh et al. 2019). Indeed, potential connectivity among lobsters in different parts of their range has been demonstrated or inferred for clawed (Hill 1990; Katz et al. 1994; Harding et al. 2005; Xue et al. 2008; Chassé and Miler 2010; Quinn et al. 2017; Gendron et al. 2019), spiny (e.g., Kough et al. 2013; Chiswell and Booth 2017; Caputi et al. 2018; Medel et al. 2018; Segura-Garciá et al. 2019), slipper (Sekiguchi and Inoue 2002; Genis-Armero et al. 2017), and squat (Gómez-Gutiérrez and Sánchez-Ortíz 1997) lobsters. All of these larval factors can also be impacted by climate change, invasive species, pollution, and other changes in the marine environment (Ennis 1995; Phillips et al. 2006). Therefore, an understanding of larval biology and ecology, particular factors impacting larval dispersal and post-larval recruitment, is important to predicting future changes to lobster populations.

The present chapter gives an overview of the larvae of different groups of lobsters. The numbers and types of larval stages and phases through which larvae develop in each taxon is first given, and then the morphology and developmental progression within and among these larval stages is described and related among groups. The duration of larval life and larval behaviors while in the water column potentially impacting their dispersal and settlement to benthic habitats are then discussed. Lastly, avenues for future research based on this overview are outlined.

2. OVERVIEW OF LOBSTER LARVAL DEVELOPMENT

2.1. General Phases of Development

This chapter follows the terminology of Martin et al. (2014), Olesen (2018), and Strathmann (2018) in distinguishing among larval phases, stages, and instars, and also in the names applied to them for the Decapoda overall. In Crustacea, different larval phases, stages, and instars are all separated from one another by a moult, but the degree of morphological and behavioral changes between them varies (Williamson 1982; Gore 1985; Anger 2001). Phases are dramatically different in many taxa (e.g., the nauplius and cypris phases of the barnacle life cycle; Olesen 2018), with the moult from one phase into another representing a putative 'metamorphosis' and a significant developmental 'jump' (Olesen 2018; Strathmann 2018). Stages differ slightly in the exact number and form of their appendages, and thus moults between stages represent anamorphic (gradual) progressions in development within the same phase (Martin et al. 2014). Instars differ only in size, with little to no developmental progression occurring between them except, maybe, very small, gradual advancements in development in a few taxa (e.g., Acheleta) (Phillips and Sastry 1980; Martin et al. 2014).

In the above sense, all lobster taxa for which the larvae are known and free-living can be said to develop through 2-3 larval phases (Williamson 1982). A number of ancestral larval phases and stages, including a nauplius-like phase ('egg-nauplius'), are presumed to be elapsed within the eggs of pleocyematan decapods like lobsters, and are thus considered to have been 'embryonized' (Martin 2014a, b; Jirkowsky et al. 2015). The main larval phase of lobsters is the zoea or mysis (Williamson 1982) phase, which follows hatching from the egg and comprises one to many stages in many taxa (and sometimes several instars). The zoea is usually planktonic or pelagic, and is presumed to be the main dispersive phase in the lobster life cycle (Phillips et al. 2006; Martin 2014b). Before this, a short-lived 'pre-zoea' phase is occasionally observed to hatch from the egg, and then moult into a first-stage zoea (Phillips and Sastry 1980; Goy 2014a; Martin 2014c; Palero et al. 2014a; Pohle and Santana 2014); however, it remains unclear whether this phase is a normal part of lobster life cycles or is instead an artefact of laboratory rearing (Phillips et al. 2006). The zoea is followed by the decapodid or 'postlarva' phase (many different names for this phase are used for particular taxa, see below and Williamson 1982; Gore 1985; Martin et al. 2014), which usually only contains one stage and is the stage that makes the transition from the water column to the benthos; after settlement, the decapodid moults into the first juvenile instar.

Among the Astacidea (clawed lobsters and crayfish), the most well-studied and commercially important species are the nephropoid (Nephropoidea) American, European, and Norway lobsters *Homarus americanus*, *H. gammarus*, and *Nephrops norvegicus*, respectively, and these species all normally develop through 3 zoeal stages (termed stages I, II, and III) and 1 decapodid ('postlarva' or stage IV lobster) (Charmantier et al. 1991; Ennis 1995; Bell et al. 2013; Johnson and Johnson 2013; Wahle et al. 2013; Goy 2014a). Under abnormal conditions, *Homarus* spp. have also been reported to moult from the last zoeal stage into 1-3 stages 'intermediate' between the zoea and decapodid (IVa, IV', and V'), for which settlement is presumed to be delayed (Charmantier and Aiken 1987; Charmantier et al. 1991). Although their role in nature, if any, is not well-understood, these 'intermediate' stages may eventually moult into juveniles and replace the decapodid phase. There are many other species of nephropoid lobsters, including species of *Metanephrops*, *Nephropsis*, *Homarinus*, and *Thaumastocheles*, but they have been less well-studied and less is known about their life histories. Some are being tested in developing fisheries, however, so they have begun to be studied more (Wahle et al. 2012; Bell et al. 2013; Goy 2014a; Heasman and Jeffs 2019). Among these other nephropoid lobsters, many inhabit deeper waters, and as an apparent consequence of this their larval stages have been reduced or suppressed. Among species of *Metanephrops*, for example, only 2 (*M. thomsoni* and *M. sagamiensis*; Uchida and Dotsu 1973; Iwata et al. 1992), 1 (*M. challenger*; Wear 1976; Heasman and Jeffs 2019), or apparently no (*M. japonicas*; Okamoto 2008) zoeal stages occur (Goy 2014a). The reef lobsters (Enoplometopoidea) are known to develop through at least 8 zoeal stages, and probably more (Iwata et al. 1991; Abrunhosa et al. 2007; Goy 2014a). The freshwater crayfishes (Astacoidea and Parastacoidea) completely lack a zoeal phase, but do develop through several distinct juvenile (or perhaps decapodid) instars post-hatch, in a process that adds posterior limbs similarly to that seen in the zoeal development of other decapods (Gherardi et al. 2010; Goy 2014a). Within the Astacidea, then, there is an apparent progression from a long and gradual development involving many zoeal stages in the Enoplometopoidea, to a more abbreviated development involving few zoeal stages in the Nephropoidea, and finally to complete loss of a true larval phase in the crayfishes (Goy 2014a).

The larvae of the Glypheidea have not yet been observed in any study (Martin 2014b). These primitive, clawed lobster-like decapods presumably do develop through free-living larval stages, perhaps resembling the zoeae of Astacidea, although the two surviving species in this group are deep-water animals (Forest 2006; Wahle et al. 2012), and thus may have an abbreviated or embryonized larval development like some nephropoids (Goy 2014a). Extinct species may have had longer or more larval stages, but fossilized glypheid larvae have also not been found or unequivocally identified (Forest 2006). A comparison of glypheid larval development and ecology to that of other lobsters and decapods must wait until further studies are done on this group.

Among the Achelata, there is a very distinct and highly modified zoeal phase in the life cycle termed the phyllosoma, which is a unique, shared character of all members of this group (Sekiguchi 1988; Lavalli and Spanier 2010; Palero et al. 2014a). Phyllosoma development is extremely gradual and variable, making it difficult to define specific stages. Based on the occurrence of specific development landmarks, typically 7-13 phyllosoma stages are defined (Phillips and Sastry 1980; Matsuda and Yamakawa 1997; Goldstein et al. 2008; Palero et al. 2014a), although in many species many more (up to 25) moults have been observed (Johnson and Knight 1966), perhaps between instars, and developmental delay by mark-time moulting has also been reported (Gore 1985); therefore, there are likely many phyllosoma instars, with the number varying considerably among and within species. The final phyllosoma stage or instar then undergoes a dramatic transformation (metamorphosis) into the decapodid phase, which is called a puerulus for spiny and furry lobsters (Palinuridae) or a nisto for slipper lobsters (Scyllaridae); there is only one puerulus or nisto stage in all taxa for which this phase has been observed, although it is worth noting that for several achelate taxa, including many slipper lobsters and all furry lobsters (*Palinurellus* spp. and *Palibythus magnificus*;

Holthuis 1991), these phases have never been reported (Phillips and Sastry 1980; Sekiguchi et al. 2007; Phillips et al. 2006).

The larvae of many species of 'mud lobsters' or Thalassinidea (Axiidea and Gebiidea) have not been thoroughly studied (Pohle et al. 2011; Pohle and Santana 2014). Of those species that have been studied, almost all have free-living planktonic larvae (although one direct developing species is known), which develop through 2-7 zoeal stages and at least one decapodid stage (sometimes called a 'megalopa') (Dworschak et al. 2012; Pohle and Santana 2014). The number of zoeal stages can vary frequently within the same species, especially for species of Callianassidae (Axiidea) (Pohle and Santana 2014), indicating that larval development among these taxa may be highly plastic.

The larvae of Polychelida are termed 'eryoneicus', and are very large, highly modified, and relatively long-lived (Martin 2014c). Because these lobsters are deep-sea inhabitants (Lavalli and Spanier 2010), relatively little is known about their larval development in detail, including what and how many stages they develop through (but see Williamson 1983; Guerao and Abelló 1996; Torres et al. 2013, 2014; Yanagimoto et al. 2015). The larval period is broadly divided into an 'eryoneicus zoea' phase, which is followed by an 'eryoneicus megalopa' (decapodid) phase; this division is based on their size and which appendages are involved in swimming (see below), but each phase presumably contains multiple stages and/or instars of similar morphology (Martin 2014c).

Squat lobsters (Anomura: Galatheoidea and Chirostyloidea), like other anomuran crabs, develop through several zoeal stages and one decapodid stage (termed a 'glaucothoe') (Tudge et al. 2012; Harvey et al. 2014). Most species of Galatheoidea for which larval development has been studied in detail develop through 4 or 5 zoeal stages, although in a few species with reduced larval phases 2-3 stages have instead been observed (Harvey et al. 2014). However, the number of zoeal stages can vary within the same species, as is common among Anomura (Boyd and Johnson 1963; Baba et al. 2011) and may be a form of settlement delay influenced by temperature, food, or other factors. For the Chirostyloidea, the exact number of larval stages is only known for one species, *Chirostylus stellaris*, which has 2 short zoeal stages (Fujita and Clark 2010), while for other species in this group similarly abbreviated development is expected based on their morphology and feeding characteristics (Harvey et al. 2014), as well as the fact that many (e.g., the yeti crabs, *Kiwa* spp.) occupy deep-sea habitats (Lovrich et al. 2011; Tudge et al. 2012).

2.2. The 'Pre-Zoea' Phase

'Pre-larva' or 'pre-zoea' stages (Figure 2A-F) have been observed in nearly all lobster groups, as well as in other decapod infraorders (stenopodid shrimps and brachyuran crabs, but not in caridean shrimps; Williamson 1982; Martin 2014a, b, d; Goy 2014b). In nephropoid (Figure 2A) and enoplometopoid (Figure 2B) lobsters, this stage emerges from the outer egg membrane, but is still covered by another thin cuticular membrane that bends the rostrum and abdominal spines downward, and prevents the appendages from moving (Goy 2014a). In *Homarus* spp., it only lasts a few hours, is commonly referred to as a 'prezoa' or hatchling, and is the standard against which embryonic eye size is compared to calculated egg development (Perkins 1972). In thalassinids, the pre-zoea stage (Figure 2C) is also very short-lived and considered to have limited to no mobility when it has been observed (Pohle and Santana 2014). It is covered by a cuticular membrane that bends the rostrum and covers the appendages, and has a rounded cephalothorax that lacks the spines and associated setae that later emerge in the zoeal phase (Pohle and Santana 2014). In polychelid lobsters (Figure 2D), a non-motile pre-zoea stage was observed with poorly developed appendages (pleopods and uropods) lacking natatory setae and covered with a cuticular membrane, and with the first two pereiopods being chelate (Williamson 1983; Guerao and Abelló 1996; Martin 2014c). Interestingly, while later-stage polychelid larvae are remarkably large (see above and below), the pre-zoeae that have been reported were very small (1 mm carapace diameter; Martin 2014c). Among anomurans, including squat lobsters, pre-zoeae have been observed in most taxa (but mainly among porcelain crabs), with the body and appendages entirely covered with a three-layered cuticular membrane; these stages were non-feeding and non-motile when observed, lasting only a few minutes to hours (Harvey et al. 2014). In achelate lobsters, the pre-zoeal stage is called a 'naupliosoma', and lasts only a few minutes to hours (Phillips and Satry 1980; Phillips et al. 2006; Palero et al. 2014a). The naupliosoma has large, biramous antennae with plumose setae (Figure 2E, F) that are apparently ready to swim, while all other appendages are coiled and non-functional (Palero et al. 2014a). Interestingly, this stage has only been observed in achelate species in which the first-stage phyllosoma also has biramous antennae (e.g., *Jasus*, *Ibacus*, and *Scyllarides*), and never within the more highly derived palinurids (e.g., the Stridentes; Parker 1884) (Palero et al. 2014a).

All lobster pre-zoeae observed thus far only remained in this stage very briefly before moulting into the first zoeal (mysis) stage, and were usually seen in a laboratory or hatchery setting. With the exception of the achelate naupliosoma, in which the antennae are free and apparently natatory, all of these stages are also obviously non-feeding and nearly immobile: the appendages are bound by the cuticular membrane, so movement, if it occurs, is only possible by flexing the body (e.g., abdomen). In stenopodids and brachyurans, the appendages of 'pre-zoeae' are also bound, and movement can only occur by body flexion (Martin 2014d; Goy 2014b). The wide occurrence of this phase suggests that it may be a real, shared component of decapod development, but its exact role remains unclear. It may, for example, correspond to the last protozoea stage of dendrobranchiate prawns (Williamson 1982; Martin 2014b; Jirkowski et al. 2015). Since it is not observed in the plankton, it might also be simply an artefact of laboratory rearing of lobster embryos (Perkins 1972; Phillips et al. 2006), although its short duration and lack of swimming ability would limit its planktonic life if it occurred in nature anyway (Phillips and Sastry 1980). Alternatively, this stage may function as a means of hatch time control. When physiologically ready to hatch, decapod eggs may hatch as pre-zoeae, which remain attached to the female's pleopods but are ready to moult when they receive some signal (female behaviors or pheromones, time of day, etc.) into the first zoea, at which point they disperse. In this way, the hatchlings can avoid being released from the female until the appropriate time, such as at night when visual predation is less likely (Phillips and Sastry 1980; Ennis 1995). The differences between the achelate naupliosoma and other lobsters' pre-zoeae suggests that in this group this stage has been adapted for some other function, but what this is remains unclear; its apparent absence in the Stridentes also begs an explanation if real. No previous study has focused on the evolutionary and ecological implications of the pre-zoea phase, but this topic may be worth investigating.

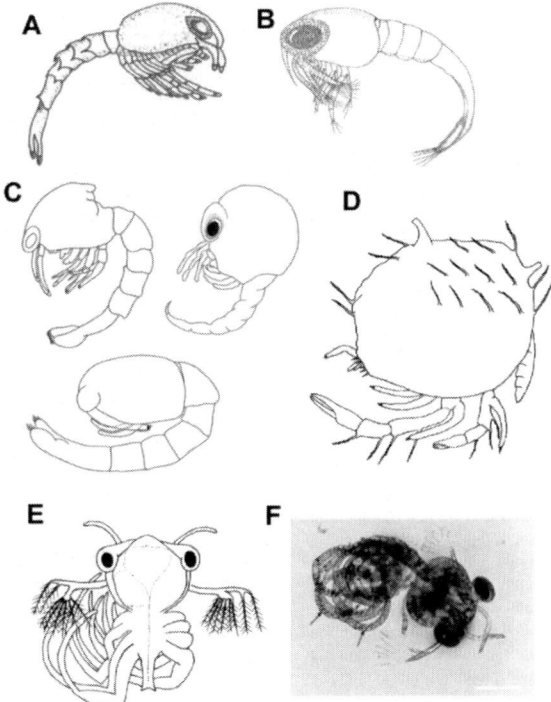

Figure 2. Examples of the 'pre-zoea' phase of various lobster taxa: (A) *Homarus americanus* (Astacidea: Nephropoidea); (B) *Enoplometopus antillensis* (Astacidea: Enoplometopoidea); (C) various calianassids (Gebiidea (formerly 'Thalassinidea'); (D) *Polycheles typhlops* (Polychelida); (E) *Jasus lalandii* (Achelata: Palinuridae); (F) *Jasus frontalis* (Achelata: Palinuridae). Images modified from: (A) Charmantier et al. (1991); Abrunhosa et al. (2007); (C) Pohle and Santana (2014); (D) Guerao and Abelló (1996); (E) Williamson and Rice (1966); (F) Dupré and Guisado (1996).

2.3. The Zoea and Decapodid Phases

Nephropoid zoeae (Figure 3A) are relatively large compared to the larvae of most other decapods, and their development is generally abbreviated in terms of the larval duration and number of moults (Gore 1985; Goy 2014a). In *Homarus* spp. and *Nephrops norvegicus*, the zoeae possess abdominal spines (which are especially remarkable in *N. norvegicus*) and natatory exopodites on the pereiopods that have plumose setae (Ennis 1995; Bell et al. 2013; Wahle et al. 2013; Goy 2014a). In all larval stages, the thoracic appendages (pereiopods and maxillipeds) are used for feeding (Phillips and Sastry 1980; Ennis 1995). The first zoea lacks pleopods and uropods, which then first appear in the second and third zoeal stages, respectively, although the pleopods do not function in swimming during the zoeal phase (Charmantier and Aiken 1987; Charmantier et al. 1991; Ennis 1995; Goy 2014a). In other nephropoid genera with even more abbreviated/advanced development, the appearances of these appendages occurs earlier (Goy 2014a; Heasman and Jeffs 2019). The first pair of pereiopods is already modified into a distinct pair of chelae at hatch, but these become especially pronounced in size from the third nephropoid zoeal stage onwards. When the third zoeal stage moults to become the decapodid (stage IV; Figure 4A), a number of changes occur in the body proportions and structures that make the decapodid much more strongly resemble the adult lobster than the larvae (Charmantier et al. 1991). Most notably, the antennae become especially long, the pereiopods lose their natatory exopodites, and the pleopods become functional, meaning that locomotion switches from the thoracic to abdominal appendages (Charmantier and Aiken 1987; Ennis 1995). This also results in a distinct improvement in horizontal swimming ability, which likely assists in settlement (Ennis 1995; Phillips et al. 2006; Stanley et al. 2016).

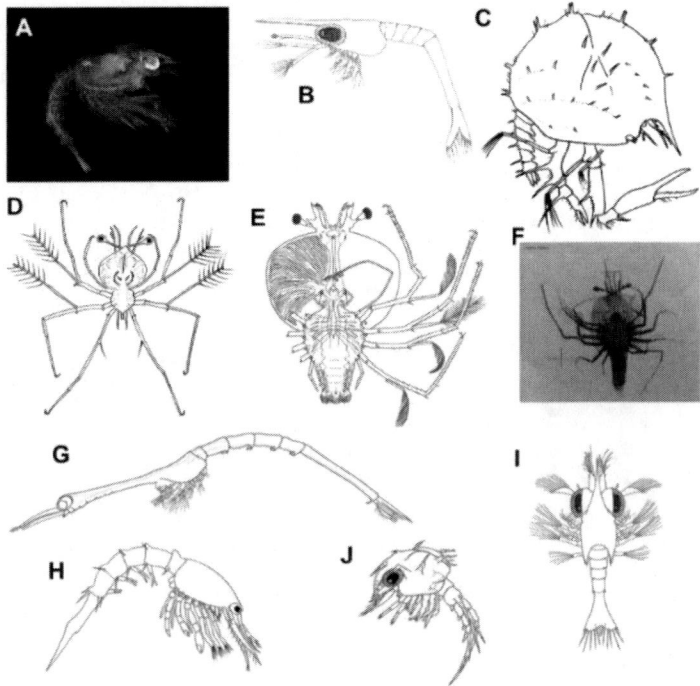

Figure 3. Examples of the zoea stages of various lobster taxa: (A) *Homarus gammarus* (Astacidea: Nephropoidea) zoea I; (B) *Enoplometopus antillensis* (Astacidea: Enoplometopoidea) zoea I; (C) *Stereomastis sculpta* (Polychelida) early stage 'eryoneicus zoea'; (D) *Jasus lalandii* (Achelata: Palinuridae) early stage phyllosoma; (E) *Scyllarus depressus* (Achelata: Scyllaridae) later stage phyllosoma; (F) *Palinurellus wieneckii* [*?Phyllamphion elegans*] (Achelata: Palinuridae (formerly 'Synaxidae')) later stage phyllosoma; (G) *Laomedia astacina* (Gebiidea: Laomediidae (formerly 'Thalassinidea')) zoea V; (H) *Axius serratus* (Axiidea (formerly 'Thalassinidea')) zoea I; (I) *Galathea intermedia* (Anomura: Galatheoidea) zoea II; (J) *Uroptychus* cf. *politus* (Anomura: Chirostyloidea) zoea I. Images modified or obtained from: (A) Wikimedia Commons (credit: H. Hillewaert, license CC-BY-SA 4.0); (B) Abrunhosa et al. (2007); (C, D) Williamson and Rice (1966); (E) Robertson (1971); (F) Smithsonian Museum of Natural History (http://n2t.net/ark:/65665/3d4c2d0a2-b644-40fb-bc30-541727f1d536); (G) Fukuda (1982); (H) Pohle et al. (2011); (I) Christiansen and Anger (1990); (J) Pike and Wear (1969).

Larval development in enoplometopoid (reef) lobsters is more gradual than that in their nephropid relatives. Hatchlings (first-stage zoeae) lack pereiopods, in addition to pleopods and uropods (Goy 2014a; Figure 3B). The pereiopods begin to appear in the second or third zoeal stage, but may not be fully functional until the fourth stage, in *Enoplometopus* spp. (Goy 2014a), while the pleopods and uropods appear in later stages, perhaps in the eighth zoeal stage or later (Abrunhosa et al. 2007; Goy 2014a). Hatchlings in this group also seem to lack abdominal spines (Goy 2014a).

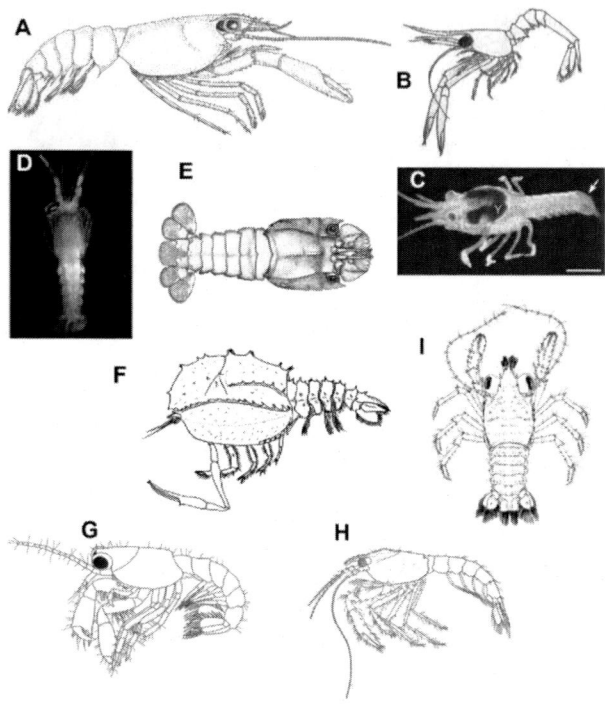

Figure 4. Examples of the decapodid phase of various lobster taxa: (A) *Homarus americanus* (Astacidea: Nephropoidea) stage IV; (B) *Enoplometopus* sp. (Astacidea: Enoplometopoidea) decapodid; (C) *Procambarus fallax* [Marmokrebs] (Astacidea: Astacoidea) juvenile instar II; (D) *Panulirus interruptus* (Achelata: Palinuridae) puerulus; (E) *Scyllarus depressus* (Achelata: Scyllaridae) nisto; (F) *Stereomastis sculpta* (Polychelida) 'eryoneicus megalopa'; (G) *Upogebia darwini* (Gebiidea (formerly 'Thalassinidae')) megalopa; (H) *Axius* sp. (Axiidea (formerly 'Thalassinidea')) megalopa; (I) *Sadayoshia edwardsii* (Anomura: Galatheoidea) glaucothoe. Images from: (A) Charmantier et al. (1991); (B) Gurney (1942); (C) Vogt and Tolley (2004); (D) Scripps Institute of Oceanography (https://scripps.ucsd.edu/zooplanktonguide/species/panulirus-interruptus); (E) Robertson (1971); (F) Williamson and Rice (1966); (G) Ngoc-Ho (1977); (H) Kurata (1965); (I) Fujita and Shokita (2005).

Enoplometopoid lobster larvae also possess a distinct process on the telson, called the 'anomuran hair' or 'thalassinidean hair', which is also found in the larvae of some thalassinid lobsters, anomuran squat lobsters, and also stenopodidean shrimps (Goy 2014a). Like in nephropids, by the time reef lobsters moult to the decapodid stage (Figure 4B), they possess the full complement of juvenile appendages, including chelae, although the more posterior pairs of pereiopods remain in their 'larval' form until later stages, and swimming switches to being abdominally propelled (Goy 2014a). Among freshwater crayfishes, there are no free-living larval stages, and juveniles hatch from the egg into a form perhaps equivalent to a nephropoid decapodid (Goy 2014a), which remains with the mother (Figure 4C). It is noteworthy, however, that the first few post-hatch stages in crayfish undergo some similar changes to those seen in the larvae of other decapods (Gore 1985). For example, in crayfish hatchlings the uropods do not appear until the second stage post-hatch, and the pleopods do not appear until the fourth stage (Goy 2014a). Pre-hatch stages in crayfishes also develop through embryonized stages that resemble free-living zoea larvae of other astacidean lobsters (Alwes and Scholtz 2006; Gherardi et al. 2010). Freshwater crayfish hatchlings also possess a unique filament, called a telson thread, which connects them to the mother's body for a time after hatching (Goy 2014a); they also possess unique hooks on the chelipeds (Astacoidea) or pereiopods (Parastacoidea).

Among the 'Thalassinidea', two distinct larval types have long been recognized, which are now known to belong to members of different decapod infraorders; specifically, the zoea larvae of the Axiidea resemble those of homarid lobsters and were thus traditionally placed in the 'homarine group' (Figure 3H), while those of the Gebiidea resemble anomuran zoeae and were placed in the 'anomuran group' (Figure 3G) (Gurney 1942; Pohle and Santana 2014). Compared with the zoeae of nephropoid lobsters, those of thalassinids are relatively more elongated and 'shrimp-like', especially those belonging to the Gebiidea; the Laomediidae within this group are especially unique in their appearance (Pohle and Santana 2014; Figure 3G).

The zoeae of both groups swim using natatory setae on the exopodites of the thoracic limbs (pereiopods). Abdominal spines like those of homarid lobster zoeae are present in some species, especially within the Axiidea (Figure 3H), but absent in others (Dworschak et al. 2012; Pohle and Santana 2014). Carapace spines may also occur in the larvae of Axiidea, but not Gebiidea (Pohle and Santana 2014). The zoeae of both groups also possess the telsonic process known as the 'thalassinidean hair', described above. As in other decapod taxa, more posterior appendages appear later in development, but the patterns in limb development vary widely among species of 'thalassinids.' In taxa with more advanced development (fewer zoeal stages), the pereiopods are present from the hatchling stage onward, whereas in those with more gradual/extended development (more zoeal stages) pairs of pereiopods may not appear until later stages (Pohle and Santana 2014). Similar variation occurs in pleopod development, while the uropods usually appear in later zoeal stages (Pohle and Santana 2014). Additional details concerning the morphological and developmental diversity seem among 'thalassinids' can be found in Pohle and Santana (2014). The decapodid ('megalopa') phase resembles the adult (Figure 4G, H), with well-developed chelipeds and abdominal appendages that are used in locomotion rather than the thoracic appendages, and with the loss of larval features like abdominal spines (Dworschak et al. 2012; Pohle and Santana 2014). Within the axiid family Calianassidae, the decapodid possesses two well-developed pairs of chelipeds, rather than just one (Pohle and Santana 2014). In some species, there is a very gradual transition from the first settling decapodid stage into subsequent juvenile forms, so there is sometimes said to be more than one decapodid stage in some species (Pohle and Santana 2014).

The phyllosoma larvae of achelate lobsters represent highly modified zoeal stages (Williamson 1982; Figure 3D-F). They have a distinct flattened, leaf-like body shape, with a greatly expanded cephalothorax and reduced abdomen, and long appendages (Sekiguchi 1988; Phillips et al. 2006; Palero et al. 2014a). The eyes are stalked and the body is mostly transparent, in contrast to other lobster larvae, which hatch with sessile eyes and develop stalked eyes in later stages and usually possess chromatohores and distinct pigmentation (e.g., thalassinids, nephropoids, and polychelids; Phillips et al. 2013a; Goy 2014a; Martin 2014c; Pohle and Santana 2014). Phyllosomas use the exopodites of the pereiopods to swim (Phillips and Sastry 1980). Early stage phyllosomas are quite small (Figure 3D), but they feed and grow extensively while in the plankton, feeding mainly on gelatinous zooplankton, and in later stages they can become quite large (Figure 3E, F), in a few cases even delaying metamorphosis to become 'giant' larvae (Sims Jr. 1964; Palero et al. 2014b). Phyllosoma morphology varies quite a lot among different achelate groups, as described in detail by Sekiguchi (1988) and Baisre (1994). The most obvious distinction is among palinurid and scyllarid phyllosomas in the shape of their antennae, as in later stage scyllarid phyllosomas these are distinctly flattened and club-shaped like those of the adults, whereas in palinurids they are more typical, whip-shaped antennae (Sekiguchi et al. 2007; Figure 3D, E). Different taxa also differ in whether they possess a rostrum in the phyllosoma stage, whether and which pereiopods are chelate, whether the cephalic shield covers the thorax (as in the controversial larval genus *Phyllamphion*; Sims Jr. 1966; Figure 3F), and so on (Sekiguchi 1988). Interestingly, one feature that is used to distinguish among major groups of adult achelate lobsters does not differ in the larvae. Specifically, while the possession of a stridualtory apparatus on the antennae is a feature used to distinguish among major groups of palinurids (Stridentes vs. 'Silentes'; Parker 1884), a recent study found precursors of this structure in the phyllosomas of not only stridente palinurids, but also silente ones and scyllarids (Fornshell and Tesei 2017). The relative sizes of the different body segments and the setation of the limbs gradually shift as development progresses, but with no major changes in morphology between phyllosoma stages/instars (Goldstein et al. 2008; Palero et al. 2014a). In the final phyllosoma stage, distinct gill buds appear on the cephalic shield (Sekiguchi 1988), after which a dramatic metamorphosis to the decapodid (puerulus or nisto) occurs (Phillips et al. 2006; Sekiguchi et al. 2007). Achelate decapodids closely resemble miniature versions of the adult lobsters of their respective groups (Figure 4D, E), and like other lobster decapodids swim quite strongly using their abdominal appendages

(Phillips et al. 2006; Kough et al. 2014; Drake et al. 2018). The palinurid puerulus (Figure 4D) is almost completely transparent and is presumed to be a non-feeding stage in most species (as it lacks a digestive gland); it is presumed to swim in search of settlement habitat as quickly as possible, and after settling moults into a feeding juvenile with pigment and a digestive gland (Phillips et al. 2006). Only a few scyllarids' nisto stages (Figure 4E) have been described, but these are also mostly unpigmented and presumed to search for settlement habitat; however, the nisto is believed to spend some time resting and hiding on the benthos between swimming bouts before it finally settles and moults to become a benthic juvenile (Sekiguchi et al. 2007).

The 'eryoneicus' larvae of polychelid lobsters (Figure 3C, 44F) are quite large (in later stages reaching 6 cm carapace length, 8 cm total length; Martin 2014c) and distinct relative to those of other lobsters. The cephalothorax is distinctly enlarged and inflated, is always larger than the abdomen, and bears a series of spines, piliers, and carinas unlike those of any other decapod larva (Williamson and Rice 1966; Martin 2014c). The larvae lack eyes, like the adults, and the rostrum projects beyond the antennules (Lavalli and Spanier 2010; Martin 2014c). The early stages ('zoeae'; Figure 3C) swim using thoracic exopodites, while in later stages swimming is accomplished by the abdominal pleopods ('megalopae'; Figure 4F) (Williamson 1983; Guerao and Abelló 1996; Martin 2014c). In the earliest stages, at least the first two pairs of pereiopods are present and chelate, although in some species more limbs may be present initially; more posterior limbs, including pleopods, are presumably added and/or become more functional as development progresses, but very few detailed studies of the larval development of polychelids have been done, so this is uncertain (Martin 2014c). Late-stage larvae ('eryoneicus megalopae') possess a full complement of five pairs of chelate pereiopods, pleopods, and uropods (Martin 2014c). Male reproductive structures have been observed among the pleopods of late-stage 'eryoneicus megalopae', so there is possibly an extremely gradual transition from the larval to the juvenile/adult phase(s) in this group, and thus they may lack a true transitional decapodid phase like that seen in other lobster and decapod taxa (Williamson 1983; Gore 1985; Martin 2014c; Olesen 2016).

The zoeae of squat lobsters (Galatheoidea and Chirostyloidea), like those of other anomuran crabs, are generally elongated and 'shrimp-like' in form, with an ovoid or cylindrical carapace (Baba et al. 2011; Harvey et al. 2014; Figure 3I, J). However, there is considerable morphological and developmental diversity within these groups. Some possess abdominal spines, and others possess extremely long and distinctive carapace spines, including posterior spines and an elongated rostrum with spines, which in some cases can be 2-3 times as long as the carapace (Baba et al. 2011; Harvey et al. 2014). While relatively little is known about the larvae of Chirostyloidea (Figure 3J), among Galatheoidea (Figure 3I) the telson can be bifurcate and armed with setae, or trapezoidal (Harvey et al. 2014). As in other taxa, zoeal locomotion is accomplished by natatory expodites on the pereiopods, and limbs are added from anterior to posterior over the course of development, although with marked variation among taxa (Baba et al. 2011; Harvey et al. 2014). In some species, early stage zoeae lack most or all pereiopods, which are then progressively added in later stages (Baba et al. 2011). Uropod buds appear in the third zoeal stage and develop further in the fourth stage, and pleopod buds generally appear in the last zoeal stage, but these limbs do not become functional until the decapodid phase (Baba et al. 2011; Harvey et al. 2014). In many anomurans, including squat lobsters, the decapodid is called a 'glaucothoe' (Gore 1985). It resembles a small crab, but with a much larger abdomen relative to the rest of its body (Figure 4I). The pleopods are used for locomotion in this phase, although the glaucothoe can walk on the pereiopods if it rests on the benthos (Harvey et al. 2014). In most squat lobsters there is one decapodid stage, although in one with very advanced development (*Munidopsis polymorpha*) the glaucothoe phase may be skipped, and the second zoeal stage moults directly into the first crab (juvenile) stage (Wilkens et al. 1990; Harvey et al. 2014).

3. PELAGIC LARVAL DURATION

The pelagic larval duration (PLD) differs considerably among lobster species, both within and among major taxonomic groups. PLD can be inferred based on a number of different methods, but the most common of these is through the rearing of larvae in a hatchery or laboratory setting. However, estimates of PLD can also be made based on field observations, including based on (1) the time elapsed between the appearance of early and late-stage larvae in plankton samples within the same locality (e.g., Annis et al. 2007), or (2) the time elapsed between peak larval hatch (estimated by sampling ovigerous ('berried') females) and peak settlement intensity (e.g., Chiswell and Booth 2017). PLD will also be affected by water temperature, as this impacts larval development rate (Ennis 1995; Anger 2001; Quinn 2019), so if a species occupies a very large geographic range its PLD in nature could differ considerably through space and time as a result of its larvae experiencing different thermal regimes.

Among clawed lobsters, the development of *Homarus americanus* in the laboratory has previously been reported to require 10.5-120.9 days from hatch to the end of stage III at temperatures of 6.7-26.3°C (Templeman 1936; Ford et al. 1979; Quinn 2019), although plankton sampling in one region suggested that development in the field may actually be 60% faster than this (Annis et al. 2007). The duration of stage IV in this species appears to be much longer (e.g., from 14-54 days at 10-22°C, representing ca. 50% of development from hatch to stage V in the absence of substrate; Templeman 1936), but more variable, possibly because it can be shortened or lengthened depending on the availability of settlement substrate (Dinning 2014) or other factors causing settlement delay or abnormal development (Charmantier and Aiken 1987; Ennis 1995; Lawton and Lavalli 1995). The development of *Homarus gammarus* larvae through zoeal stages I-III appears to be somewhat shorter than its congener, being 13.5-35.3 days at 14.2-22.0°C (Schmalenbach and Franke 2010). *Nephrops norvegicus* required 14-20 days (2-3 weeks) to complete stages I-III (Poulsen 1946; Phillips and Sastry 1980; Goy 2014a). The *Metanephrops* species that have been studied required only 4-9, 6-8, and 3-

4 days from hatch to the moult to decapodid (*M. thomsoni, M. sagamiensis,* and *M. challenger*, respectively; Goy 2014a and references therein) *Enoplometopus occidentalis* and *E. antillensis* were reared from hatch to the eighth zoeal stage, requiring 55 and 64-96 days, respectively, although at this point they still had not developed pleopods, so presumably would have required more zoeal stages before moulting to the decapodid (Iwata et al. 1991; Abrunhosa at al. 2007; Goy 2014a).

The duration of the phyllosoma stage of achelate lobster varies widely, from a few weeks to ca. 2 years (Phillips and Sastry 1980; Sekiguchi 1988; Phillips et al. 2006 and references therein; Chiswell and Booth 2017). In general, the scyllarid species for which PLD has been estimated seem to have shorter PLDS overall (ca. 6 weeks in *Scyllarus chacei* (Robertson 1968) to 9 months in *Scyllarides nodifer* (Sims Jr. 1965; Robertson 1969a)) than those of palinurids (ca. 4-6 months in *Panulirus homarus* (Berry 1974a) to 12-22 months in *Jasus edwardsii* (Lesser 1978; Chiswell and Booth 2017)) (Phllips and Sastry 1980). Interestingly, whereas clawed lobster larvae reared in the laboratory are thought to develop more slowly than they do in the field (see above), achelate phyllosomas may actually develop more quickly in the laboratory than in the real ocean (Phillips and Sastry 1980), perhaps due to the lack of certain oceanographic stimuli and/or the need to delay settlement in nature. The durations of the puerulus and nisto stages are less well-studied. The puerulus of *Panulirus interruptus* was estimated to have a lifetime of 2-3 months (Serfing and Ford 1975), while that of *P. argus* moulted within 5-10 days when in captivity (Ting 1973). The nisto stage of the scyllarid *Petrarctus demani* lasted 5-6 days, that of *Ibacus peroni* lasted 14-24 days, and that of *Thenus orientalis* lasted 7 days (Phillips et al. 2006 and references therein; Sekiguchi et al. 2007).

The entire larval phase of any polychelid lobster has never been reared in the laboratory, and larvae are not common in plankton samples, so we do not currently have a good understanding of these lobster's PLD. They have been estimated to last for a very long, but unknown (perhaps several months?), period based on the large size disparity between hatchlings and late-stage 'eryoneicus megalopae' and the extremely gradual developmental progression between subsequent larval stages or instars (Williamson 1983; Guerao and Abelló 1996; Martin 2014c).

Among 'thalassinids,' zoeal durations have been estimated to be as short as 2-3 days to as long as several months, but tend to be on the short side (Tamaki et al. 1996; Newman et al. 2006). *Callichirus kraussi* (Axiidea) zoeae complete development after 3-5 days (Forbes 1973). The six zoeal and one decapodid stages of *Nihonotrypaea japonica* (Axiidea) were estimated to be completed in 20-25 days at 18.5-24.5°C (Tamaki et al. 1996). *Upogebia africana* and *U. capensis* (Gebiidea) required 7-27 and 7-19 days to complete their three zoeal stages at temperatures of 12-29 and 11-26°C, respectively (Newman et al. 2006).

Among squat lobsters, the larvae of many species have not been reared throughout their development in the laboratory (Harvey et al. 2014). Of those for which the PLD is known or inferred, there is a wide range of PLDs depending on whether a species has abbreviated development, as noted above, as well as whether its larvae are lecithotrophic or planktotrophic (Baba et al. 2011). The only species of Chirostyloidea for which the PLD is know is *Chirostylus stellaris*, which completed development through its 2 zoeal stages in only 2 days (Fujita and Clarke 2010). Among the Galatheoidea, species of *Munidopsis* with lecithotrophic larvae have PLDs of 14-21 days (Wilkensen et al. 1990; Baba et al. 2011), while species with planktotrophic larvae have PLDs from 13-16 (*Galathea* spp.) to 55 days (*Pleuroncodes monodon*) (Baba et al. 2011 and references therein). The species *Pleuroncodes planipes* (Galatheoidea) completed its larval phase through four or five zoeal stages within between 60.5-127.5 days at temperatures of 12-20°C (Boyd and Johnson 1963).

4. LARVAL BEHAVIORS

4.1. Changes in Vertical Position in the Water Column

Larvae exhibit a wide array of behaviors that are important for their survival, including those involved in feeding, predator avoidance, intra-

and interspecific interactions, and so on (Phillips and Sastry 1980; Ennis 1995; Phillips et al. 2006; Pineda and Reyns 2018). Within the present chapter, behaviors that have the potential to directly impact the duration and extent of dispersal by larvae while they are in the plankton are focused upon, including primarily vertical positioning behaviors and horizontal swimming. The challenge facing all organisms with planktonic larvae is that, while some dispersal is needed to avoid crowding and inbreeding effects in benthic populations, too much dispersal increases the chance of larval mortality, particularly by dispersing away from the habitats that they will need to inhabit as juveniles and adults (Pineda and Reyns 2018). It is well-established that if marine invertebrate larvae simply behaved as passively dispersing particles, then they would disperse much further than the evidence available suggests that they do (Butler IV et al. 2011; Pineda and Reyns 2018). Therefore, larvae must use behaviors to control (and mainly reduce) their dispersal as larvae. This also applies to lobsters, as dispersal modeling studies of clawed (Katz et al. 1994; Stanley 2015) and spiny (e.g., Butler IV et al. 2011; Kough et al. 2013; Drake et al. 2018) lobsters, among others, have shown that incorporating larval behaviors into such models' simulations can drastically change the predicted dispersal trajectories and distances of larvae. Below, the known or inferred behaviors of lobster larvae that might impact their dispersal, and the demographic and ecological implications thereof, are described.

Lobster larvae generally move by means of their thoracic exopodites in the zoeal stages, and their abdominal pleopods in the decapodid phase, with tail flicks also occasionally used in all stages for rapid escape responses (Ennis 1986, 1995; Jackson and MacMillan 2000). The zoeal exopodite apparatus is quite effective at moving larvae throughout the water column, especially when their body is smaller, but has limited ability to generate horizontal thrust and resist strong currents (Ennis 1986, 1995; Stanley 2015; Pineda and Reyns 2018). Conversely, the use of pleopod swimming works better for larger-bodied larvae and allows for rapid swimming in both the horizontal and vertical planes, as well as against relatively strong currents (Phillips et al. 2006; Palero et al. 2014a; Stanley 2015). Although most decapod zoeae, including those of lobsters, are positively rheotactic and will face into and attempt to resist currents (Ennis 1986, 1995; Anger 2001), which likely does help to reduce dispersal to some extent, in most cases they cannot generate enough forward thrust with their exopodites to resist more than the weakest of currents (Ennis 1995; Pineda and Reyns 2018). Therefore, the zoeal stages more likely regulate their dispersal by changing their vertical position in the water column (Hill 1991; Pineda and Reyns 2018).

Due to Eckman, Coriolis, and friction effects, currents deeper in the water column are generally weaker and can move in different directions than those at the surface (Pineda and Reyns 2018). In coastal areas, the vertical current profile is also markedly altered by the tidal phase: on the flood tide fast surface currents are directed shoreward, while on the ebb tide they are directed seaward, and between these (at slack tide) currents are generally weaker (Hill 1991; Pineda and Reyns 2018). A larva can thus promote or resist its dispersal away from the location at which it hatched by controlling its position in the water column at the right times and in response to the appropriate cues (Hill 1991). As most commercially exploited lobsters inhabit relatively shallow-water coastal regions, it could be advantageous for their larvae to either resist dispersal altogether, or to disperse away from parental habitat in early larval stages and then behave in ways that help them return to parental habitats in later larval stages (Phillips et al. 2006).

Among the clawed lobsters, the behaviors of the larvae of the American lobster *H. americanus* are perhaps the most extensively studied (Ennis 1995). Lobster larvae are able to sense and respond to stimuli such as light, salinity, hydrostatic pressure, gravity, the smell of conspecifics, currents, and temperature, all of which could potentially serve as cues to adjust swimming and vertical positioning behavior (Hadley 1908; Boudreau et al. 1992; Ennis 1995). Increasing hydrostatic pressure and strongly decreasing temperature (and perhaps salinity) gradients eventually cause larvae to swim in a way that will prevent them from sinking and rise in the water column (Ennis 1995). Responses to light appear to vary in complex ways among and within stages. Generally, early stage I larvae swim toward bright light, but then become negatively phototactic later in this stage and remain so for most of stages II and III (Hadley 1908; Harding et al. 1987; Ennis 1995; Stanley et al. 2016). In stage IV, the decapodid is initially positively phototactic, but then starts to avoid light later in the stage, when it is believed to begin sampling the bottom in attempt to settle (Ennis 1995; Meyers et al. 2012; Barret et al. 2017). Similar findings were also found for *H. gammarus* larvae (Schmalenbach and Buchholz 2010). These responses to light in the laboratory agree with the results of some field sampling efforts, in which the distributions of stage I and IV lobsters showed diel patterns while those of stages II and III remained deeper regardless of the time of day (Harding et al. 1987; Tully and Ó Céidigh 1987). However, even larval stages that appear to avoid the light, and therefore the surface waters, seem to remain above the local thermocline and pycnocline (Harding et al. 1987; Boudreau et al. 1992; Meyers et al. 2012; Barret et al. 2017). Conversely, no obvious ontogenetic or diel changes in vertical distributions were found in regions where the water column was well-mixed and unstratified (Ennis 1995 and references therein). Homarid lobster larvae may thus inhabit the surface waters as hatchlings to promote dispersal away from their parents, but then in subsequent zoeal stages move into deeper waters to reduce dispersal, using cues from the water column stratification to avoid extremely deep and physiologically disadvantageous waters. Diel migrations in stage I could help this stage to catch prey or avoid predators, and may also involve the use of tidal cycles to initially promote and then perhaps avoid dispersal (e.g., Hill 1991; Phillips et al. 2006; Pineda and Reyns 2018), although this is only speculation at present. In the early part of stage IV, the return to surface waters may somehow help this stage to navigate back toward parental habitats, although it is not clear exactly how or if this is done (Hadley 1908; Ennis 1995; one study suggested that stage IV lobsters swim northward or shoreward: Cobb et al. 1989). The exact implications of these specific behaviors to lobster larval dispersal (e.g., whether they actually reduce larval dispersal) have not yet been tested at a large scale (although Stanley (2015) attempted to do so at a very small scale). If they do, however, it may mean that there is less connectivity among American lobster populations than has previously been inferred (e.g., Gendron et al. 2019).

Achelate lobster larvae possess a comparable range of sensitivities to environmental stimuli and cues as clawed lobsters have (Phillips and Sastry 1980; Phillips et al. 2006), in addition to a putative magnetic compass sense in palinurids (Lohrmann and Ernst 2013) and stronger swimming ability in their later phyllosoma stages, due to their large size (Phillips et al. 2006; Sekiguchi et al. 2007). The larvae of many or all achelate species also spend at least part of their pelagic life in relatively open-ocean environments, meaning that depth regulation and having some means of returning to shallow coastal parental habitats may be especially critical to them (Phillips et al. 2006). Indeed, some species inhabit the coasts of isolated islands or seamounts as adults, yet their larvae are found within large gyres in the middle of the world's ocean basins (Phillips et al. 2006; Chiswell and Booth 2017). Perhaps the larvae of such species behave in such a way that they are initially entrained in such large oceanic gyres, disperse away from home, feed and grow within these systems, and then, when the time is right, metamorphose just as the gyre brings them back toward their place of origin (Phillips et al. 2006; Chiswell and Booth 2017); however, this hypothesis has not been conclusively demonstrated yet. There is evidence from plankton sampling and laboratory studies that phyllosomas exhibit diel and/or ontogenetic changes in their vertical positioning in the water column (Sekiguchi et al. 2007; Butler IV et al. 2011). For example, the Caribbean spiny lobster *Panulirus argus* initially occupies shallower, well-lit surface waters in its early stages and only moves over a comparatively narrow range of depths, whereas in later stages the phyllosomas move to deeper depths on average, but also range over a much larger portion of the water column (Butler IV et al. 2011); further, there is evidence that earlier stages might be attracted to light and thus spend more time in the surface waters during the day than at night, but this behavior was lost in later stages. While this pattern of vertical migration may be 'typical' for achelate larvae (Sims Jr. 1965; Sekiguchi 1988; Phillips et al. 2006), there are some apparent exceptions, as for example the palinurid *Jasus edwardsii* did not show evidence of much regular vertical movement and the scyllarid *Thenus orientalis* seems to have limited ability to move to surface waters (Phillips et al. 2006 and references therein). When such changes in the vertical positioning of larvae occurred, they resulted in them dispersing less far on average than if they drifted passively in surface waters in a dispersal model of *P. argus* (Butler IV et al. 2011; Kough et al. 2013). A number of species' phyllosomas, particularly those of scyllarids, are often found associated with gelatinous zooplankton, such as jellyfish and salps (Wakabayashi and Yamada 2012). Many of these species appear to ride these zooplankton for a long time as they consume them piecemeal (Sekiguchi et al. 2007; Wakabayashi and Yamada 2012; Spanier and Lavalli 2013), and thus this novel behavior could also impact larval dispersal in ways that have not yet been quantified.

Relatively little is known about the behavior of the larvae of polychelids, in the laboratory or in nature. Adult polychelids are inhabitants of the deep sea, while their larvae are generally found higher up in the water column, so presumably vertical migrations occur upward in the earlier larval stages and then downward in the later (settling) stages. A recent study in the deep Mediterranean provided suggestive evidence of these behaviors, as they captured early stage 'eryoneicus zoeae' of *Polycheles typhlops* in plankton samples in shallower waters than where the adults live (although still below the euphotic zone, in depths of 400-600 m), and later stage 'eryoneicus zoeae' were found even shallower (at 200-400 m); conversely, decapodids ('eryoneicus megalopae') were found in deeper waters approaching adult habitats (600-800 m) (Torres et al. 2013). Presumably these larvae regulate their position based on hydrostatic pressure, and maybe also while following food sources (Torres et al. 2013). The implications of such behaviors to polychelid recruitment are unclear, but as deep-sea habitats can be sparsely distributed and polychelid larval duration is presumed to be long, perhaps these behaviors facilitate dispersal to new, scattered habitats.

Studies of the larvae of 'thalassinids' have historically been quite limited (Pohle and Santana 2014), so there are no detailed observations of their behavior in the laboratory or the field. Many thalassinid adults inhabit shallow, coastal waters and tidal estuaries, where they burrow in sediment, banks, or even rocky shorelines (Saigusa et al. 2018), so presumably their larvae undergo similar behaviors to those seen in other lobster and crab taxa to help them reduce dispersal and return to parental/juvenile habitats (Hill 1991; Ennis 1995; Phillips et al. 2006). However, the details of these behaviors remain to be elucidated, and the relatively short larval duration of many species in these groups noted above (Newman et al. 2006; Pohle and Santana 2014) may also help to reduce/avoid dispersal independently (or in concert with) larval behaviors.

The detailed larval life of most squat lobsters has not yet been studied, including their larval dispersal and behavior *in situ* (Baba et al. 2011). For the galatheoideans *Munida gregaria*, *Pleuroncodes monodon*, and *Pleuroncodes planipes*, several detailed studies have examined their distributions in the ocean (e.g., Dellatorre and Barón 2008; Gómez-Gutiérrez and Sánchez-Ortíz 1997; Roa et al. 1995; see also other studies cited by Baba et al. 2011). Seasonal differences in the distributions of these species' larval stages have been observed, with early stage larvae initially found in shallower waters closer to shore, while late-stage zoeae and postlarvae are found further from shore (Baba et al. 2011). These patterns may suggest that juveniles recruit to different benthic habitats than those of adults in these species (Baba et al. 2011). It is unclear from the studies done thus far whether these changes in larval distribution with ontogeny occur as a result of active behaviors by the larvae regulating their depth and/or horizontal displacement, or are simply the result of passive drift with local currents. Given that behaviors controlling vertical positioning and dispersal are known in other crabs and lobsters (e.g., Butler IV et al. 2011; Meyers et al. 2012; Torres et al. 2013), they likely also occur among squat lobster larvae, although this requires further study.

4.2. Horizontal Swimming, Navigation, and Settlement Behavior

The decapodid stages of Crustacea Decapoda are known as some of the strongest swimmers among invertebrate larvae (Martin 2014b; Pineda and

Reyns 2018), so there is much potential for horizontal swimming and/or navigation by lobster decapodids to greatly modify the extent and destinations of their dispersal.

Among clawed lobsters, stage IV *Homarus americanus* have previously been reported to be able to swim in flumes against currents of up to 9-15 cm/s for 10-30 min (Ennis 1986, 1995; Cobb et al. 1989), and short-term (ca. < 1 min to < 1 h) field observations by divers of released decapodids suggested that they could swim as fast as 24 cm/s, in a northern or eastern direction (Cobb et al. 1989). Swimming by stage IV lobsters at such speeds could greatly modify their dispersal, especially if they did so while navigating in a particular direction (Katz et al. 1994). However, the applicability of such laboratory-based or limited field observations to lobster larvae in nature remains unclear. Whether and how *H. americanus* decapodids could navigate toward settlement habitat remains an open question; they could certainly use a number of cues to do so, including salinity changes, conspecific odors, celestial cues, polarized light, turbulence, breaking down of thermal stratification, or others (Hadley 1908; Boudreau et al. 1992; Ennis 1995; Meyers et al. 2012; Dinning 2014; Chiasson et al. 2015; Stanley et al. 2016; Barret et al. 2017), but this requires further study. Swimming by stage IV lobsters is also almost certainly not continuous in nature, so understanding what cues induce swimming, and for how long, is also important. There is some evidence that this stage swims more in the presence of light (i.e., at daytime) (Hadley 1908; Stanley et al. 2016). Exactly how this stage swims in relation to currents is also important. By resisting currents, it could reduce dispersal, but not prevent it completely, but by swimming across currents or in the same direction as them if they are directed toward parental habitat, it could reach desirable settlement grounds (Ennis 1995). Importantly, the swimming ability of stage IV lobsters also appears to have been previously overestimated, as a recent study (Stanley et al. 2016) showed that without the ability to take advantage of boundary layer effects near the bottoms of flumes they could only maintain their position by swimming against currents of up to ~6 cm/s rather than the 9-24 cm/s speeds suggested in previous studies. This still makes them strong swimmers for invertebrates (Pineda and Reyns 2018), but the impact of their swimming on dispersal may not be as extensive as previously thought. Ultimately, swimming by homarid lobster decapodids is likely important to their dispersal, but more research is still needed to understand this behavior before its importance can be exactly quantified.

Among achelate lobsters, some horizontal swimming may be achieved by late-stage phyllosomas, since these are quite large and capable swimmers using their thoracic expodites (Phillips et al. 2006). However, the decapodid stages of these lobsters (puerulus and nisto) are likely the most important horizontally swimming stage in their life cycle, which is presumably an essentially means for them to navigate to coastal juvenile/adult habitats. The pueruli of many shallow-water palinurids swim at high speeds (6-46 cm/s; Phillips et al. 2006), enabling them to complete cross-shelf movements (Pineda and Reyns 2018). During this stage, the puerulus may exhibit a diel cycle, in which it alternates between actively swimming in the water column during the night and sheltering/hiding on the benthos during the day (Phillips et al. 2006); similar patterns have also been noted for the scyllarid nisto (Sekiguchi et al. 2007). While swimming, achelate decapodids, especially palinurid pueruli (which have been more well-studied than scyllarid nistos), tend to move continuously in straight lines, and there is evidence from field surveys that pueruli must navigate specifically toward settlement habitat rather than swimming in random directions (Phillips et al. 2006). A recent study tethering the pueruli of *Panulirus argus* in floating observation platforms (Kough et al. 2014) confirmed that they oriented themselves and consistently swam in the same direction that would take them toward coastal settlement habitats, independently of tidal and celestial cues. Adult *P. argus* are known to possess magnetoreceptors in their heads (Lohrmann and Ernst 2013) that enable them to navigate while performing short-term movements and long-term migrations, and such organs may also occur in the puerulus stage; therefore, decapodids in this achelate lobster species, and presumably also in others, appear to be able to navigate using an internal compass sense. The impacts of this behavior on larval dispersal by achelate lobsters remain to be fully explored (but see Drake et al. 2018), but it likely plays a role in reducing dispersal, increasing recruitment, and decreasing larval loss and mortality (Pineda and Reyns 2018).

As noted elsewhere, we know relatively little above the larval ecology of 'thalassinids' and polychelids in nature, including their behavior, and thus how their decapodid phases swim in a horizontal plane and whether/how this impacts their settlement and recruitment to juvenile habitats. Presumably, such behaviors exist, since many thalassinids need to return to coastal adult habitats to settle, with which horizontal swimming by decapodids could help, and late-stage polychelid larvae are quite large, and thus likely are capable swimmers. However, for polychelids, given that adults are primarily deep-water animals and that the larvae may find themselves quite high in the water column (Torres et al. 2013), perhaps vertical swimming is more important to help them achieve successful settlement.

As noted above, changes in squat lobster larval distributions are known to occur with ontogeny in some species (Baba et al. 2011), but the behaviors of the larvae responsible for these changes, if any, remain unclear. However, offshore-directed horizontal swimming by late-stage larvae and postlarvae could be responsible for their shift toward deeper water and their presumed juvenile habitats. It should also be noted that some species of squat lobsters, such as *Pleuroncodes planipes*, are pelagic as adults (Haye et al. 2010), and thus these species may disperse as much or more as adults as they do as larvae, so a complete understanding of their dispersal ecology and population demography requires an understanding of the pelagic dispersal of adults as well as larvae.

5. FUTURE RESEARCH DIRECTIONS

The larvae of many lobster taxa (particularly non-commercial species) have not been studied, or if they have been studied their detailed development, behavior, and ecology have not been examined.

Particularly lacking are observations of the complete development (along with the PLD) of reef lobsters, most thalassinids, polychelids, some squat lobsters, and furry lobsters. Whether extant glypheids have larvae, and if so what their development is like, is also completely unknown. Once more taxa are assessed, evolutionary studies may then be done in relation to developmental shifts among lobsters, for example whether abbreviated larval development and life in the deep sea are related in lobster taxa.

The biological reality and ecological function of the 'pre-zoea' stages seen in nearly all lobster taxa investigated thus far remains unclear, but the widespread occurrence of such stages among decapods suggests that they may be a real component of larval development with some important function; this should be investigated.

While some aspects of the vertical migratory and horizontal swimming behaviors of lobster larvae have been investigated, particularly in a laboratory setting, the actual behaviors of larvae and the implications of such larval behaviors in nature remain unclear. Answering these questions will require intensive field studies combined with oceanographic modeling work. Behaviors by the larvae of poorly studied taxa (polychelids, thalassinids, squat lobsters, and obviously glypheids, as well as non-commercial clawed, spiny, slipper, and reef lobsters) should also be investigated.

Lastly, thorough and empirically validated examinations of the role of larval dispersal and connectivity among lobster populations are needed to truly allow lobster fisheries' recruitment to be accurately forecasted and for lobster populations to be fished sustainably. While oceanographic dispersal models, genetic markers, and other tools have been used to infer the potential connectivity among lobster stocks, these tools' predictions or their demographic implications are rarely validated to an extent that allows them to be used to make satisfactory predictions for stock assessment and fishery management. First steps toward achieving such predictive success should involve detailed and thorough investigations of the biology and behavior of lobster larvae to fill the knowledge gaps identified above.

REFERENCES

Abrunhosa FA, Santana MWP, Pires MAB (2007) The early larval development of the tropical reef lobster *Enoplometopus antillensis* Lütken (Astacidea, Enoplometopidae) reared in the laboratory. *Rev Bras Zool* 24:383-396.

Alwes F, Scholtz G (2006) Stages and other aspects of the embryology of the parthenogenetic Marmokrebs (Decapoda, Reptantia, Astacida). *Dev Genes Evol* 216:169-184.

Anger K (2001) *Crustacean Issues 12: The Biology of Decapod Crustacean Larvae*. Rotterdam, The Netherlands: A. A. Balkema.

Annis ER, Incze LS, Wolff N, Steneck RS (2007) Estimates of *in situ* larval development time for lobster, *Homarus americanus. J Crustac Biol* 27:454-462.

Baba K, Fujita Y, Wehrtmann IS, Scholtz G (2011) Developmental biology of squat lobsters. In: *Crustacean Issues 20: The Biology of Squat Lobsters*. Poore GCB, Ahyong ST, Taylor J (eds.). Collingwood, Australia: CRC Press, p. 105-148.

Baisre JA (1994) Phyllosoma larvae and the phylogeny of Palinuroidea (Crustacea: Decapoda): A review. *Aust J Mar Freshwat Res* 45:925-944.

Barret L, Miron G, Ouellet P, Tremblay R (2017) Settlement behavior of American lobster (*Homarus americanus*): effect of female origin and developmental temperature. *Fish Oceanogr* 26:69-82.

Bell M, Tuck I, Dobby H (2013) *Nephrops* species. In: *Lobsters: Biology, Management, Aquaculture and Fisheries* (2nd Edition). Phillips B. (ed.). West Sussex, UK: Wiley-Blackwell, p. 357-413.

Berry PF (1974) Palinurid and scyllarid lobster larvae of the Natal coast, South Africa. *Oceanogr Res Inst (Durban), Invest Rep* 22:1-55.

Boudreau B, Simard Y, Bourget E (1992) Influence of a thermocline on vertical distribution and settlement of post-larvae of the American lobster, *Homarus americanus. J Exp Mar Biol Ecol* 162:35-49.

Boyd CM, Johnson MW (1963) Variations in the larval stages of a decapod crustacean, *Pleuronectes planipes* (Galatheidae). *Biol Bull* 124:141-152.

Boyko CB (2006) New and historical records of polychelid lobsters (Crustacea: Decapoda: Polychelidae) from the Yale Peabody Museum collections. *Bull Peabody Mus Nat Hist* 47:37-48.

Bracken-Grissom HD, Ahyong ST, Wilkinson RD, Feldman RM, Schweitzer CE, Breinholt JW, Bendall M, Palero F, Chan T-Y, Felder DL, Robles R, Chu K-H, Tsang L-M, Kim D, Martin JW, Crandall KA (2014) The emergence of lobsters: Phylogenetic relationships, morphological evolution and divergence time comparisons of an ancient group (Decapoda: Achelata, Astacidea, Glypheidea, Polychelida). *Syst Biol* 63:457-479.

Butler MJ IV, Goldstein JS, Matsuda H, Cowan RK (2011) Behavior constrains the dispersal of long-lived spiny lobster larvae. *Mar Ecol Prog Ser* 422:223-237.

Caputi N, Feng M, Denham A, de Lestang S, Penn J, Slawinski D, Pearce A, How J (2018) Optimizing an oceanographic-larval model for assessment of the puerulus settlement of the western rock lobster, *Panulirus cygnus*, in Western Australia. *Bull Mar Sci* 94:775-800.

Chan T-Y (2010) Annotated checklist of the world's marine lobsters (Crustacea: Decapoda: Astacidea, Glypheidea, Achelata, Polychelida). *Raffles Bull Zool* Suppl. No. 23:153-181.

Charmantier G, Aiken DE (1987) Intermediate larval and postlarval stages of *Homarus americanus* H. Milne Edwards, 1837 (Crustacea: Decapoda). *J Crustac Biol* 7:525-535.

Charmantier G, Charmantier-Daures M, Aiken DE (1991) Metamorphosis in the lobster *Homarus* (Decapoda): A review. *J Crustac Biol* 11:481–495.

Chassé J, Miller RJ (2010) Lobster larval transport in the southern Gulf of St. Lawrence. *Fish Oceanogr* 19:319-338.

Chiasson M, Miron G, Daoud D, Mallet MD (2015) Effect of temperature on the behavior of stage IV American lobster (*Homarus americanus*) larvae. *J Shellfish Res* 34:545-554.

Chiswell SM, Booth JD (2017) Evolution of long larval life in the Australasian rock lobster *Jasus edwardsii*. *Mar Ecol Prog Ser* 576:69-87.

Christiansen ME, Anger K (1990) Complete larval development of *Galathea intermedia* Lilljeborg reared in laboratory culture (Anomura: Galatheidae). *J Crustac Biol* 10:87-111.

Cobb JS, Wang D, Campbell DB, Rooney P (1989) Speed and direction of swimming by postlarvae of the American lobster. *Trans Am Fish Soc* 118:82-86.

Davie PJF (1990) A new genus and species of marine crayfish, *Palibythus magnificus*, and new records of *Palinurellus* (Decapoda: Palinuridae) from the Pacific Ocean. *Invertebr Taxon* 4:685-695.

Dellatorre FG, Barón PJ (2008) Multiple spawning and length of embryonic development of *Munida gregaria* in northern Patagonia (Argentina). *J Mar Biol Assoc UK* 88:975-981.

Dinning KM (2014) *Effect of Substrate on Settlement Behaviour, Development, Growth, and Survival of American Lobster Postlarvae, and Evidence that Mud Bottom can serve as Secondary Nursery Habitat*. M.Sc. Thesis, University of New Brunswick, Saint John, NB, Canada.

Drake PT, Edwards CA, Morgan SG, Satterthwaite EV (2018) Shoreward swimming boosts modeled nearshore larval supply and pelagic connectivity in a coastal upwelling region. *J Mar Syst* 187:96-110.

Dupré E, Guisado C (1996) Early stages of phyllosoma of the spiny lobster of Juan Fernandez (*Jasus frontalis*) maintained in laboratory conditions. *Invest Mar, Valparaíso* 24:39-50. [In Spanish, abstract in English.]

Dworschak PC, Felder DL, Tudge CC (2012) Infraorders Axiidea de Saint Laurent, 1979 and Gebiidea de Saint Laurent, 1979 (formerly known collectively as Thalassinidea). In: *Treatise on Zoology, the Crustacea. Vol. 9B*. Schram F, von Vaupel Klein JC (eds.). Leiden, The Netherlands: Brill, p. 109-219.

Ennis GP (1986) Swimming ability of larval American lobsters, *Homarus americanus*, in flowing water. *Can J Fish Aquat Sci* 43:2177-2183.

Ennis GP (1995) Larval and postlarval ecology. In: *Biology of the Lobster Homarus americanus*. Factor JR (ed.). New York, NY, USA: Academic Press, Inc., p. 23-46.

Forbes AT (1973) An unusual abbreviated larval life history in the estuarine burrowing prawn *Callianassa kraussi* (Crustacea: Decapoda: Thalassinidea). *Mar Biol* 22:361-365.

Ford RF, Felix JR, Johnson RL, Carlberg JM, Van Olster JC (1979) Effects of fluctuating and constant temperatures and chemicals in thermal effluent on growth and survival of the American lobster (*Homarus americanus*). *Proc World Maricul Soc* 10:139-158.

Forest J (2006) The recent glypheids and their relationship with their fossil relatives (Decapoda, Reptantia). *Crustaceana* 79:795-820.

Fornshell JA, Tesei A (2017) Observations on the sound producing organs in achelate lobster larvae. *Arthropods* 6:36-46.

Fujita Y, Clark PF (2010) The larval development of *Chirostylus stellaris* Osawa, 2007 (Decapoda: Anomura: Chirostylidae) described from laboratory-reared material. *Crustac Res* 39:53-64.

Fujta Y, Shokita S (2005) The complete larval development of *Sadayoshia edwarsii* (37) (Decapoda: Anomura: Galatheidae) described from laboratory-reared material. *J Nat Hist* 39:865-886.

Fukuda Y (1982) Zoeal stages of the burrowing mud shrimp *Laomedia astacina* de Haan (Decapoda: Thalassinidea: Laomediidae) reared in the laboratory. *Proc Japan Soc System Zool* 24:19-31.

Gendron L, Lefaivre D, Sainte-Marie B (2019) Local egg production and larval losses to advection contribute to interannual and long-term variability of American lobster (*Homarus americanus*) settlement intensity. *Can J Fish Aquat Sci* 76:350-363.

Genis-Armero R, Guerao G, Abelló P, González-Gordillo JI, Cuesta JA, Corbari L, Clark PF, Capaccioni-Azzati R, Palero F (2017) Possible amphi-Atlantic dispersal of *Scyllarus* lobsters (Crustacea: Scyllaridae): molecular and larval evidence. *Zootaxa* 4306:325-338.

Gherardi F, Souty-Grosset C, Vogt G, Diéguez-Uribeondo J, Crandall KA (2010) Infraorder Astacidea Latreille, 1802 p.p.: The freshwater crayfish. In: *Treatise on Zoology, the Crustacea. Vol. 9A*. Schram F, von Vaupel Klein JC (eds.). Leiden, The Netherlands: Brill, p. 269-423.

Goldstein JS, Matsuda H, Takenouchi T, Butler MJ IV (2008) The complete development of larval Caribbean spiny lobster *Panulirus argus* (Latreille, 1804) in culture. *J Crustac Biol* 28:306-327.

Gómez-Gutiérrez J, Sánchez-Ortíz CA (1997) Larval drift and population structure of the pelagic phase of *Pleuroncodes planipes* (Stimpson) (Crustacea: Galatheidae) off the southwest coast of Baja California, México. *Bull Mar Sci* 61:305-325.

Gore (1985) Moulting and growth in decapod larvae. In: *Crustacean Issues 2: Larval Growth*. Wenner AM (ed.). Rotterdam, The Netherlands: A. A. Balkema, p. 1-65.

Goy JW (2014a) Astacidea. In: *Atlas of Crustacean Larvae*. Martin JW, Olesen J, Høeg JT (eds.). Baltimore, MD, USA: Johns Hopkins University Press, p. 256-262.

Goy JW (2014b) Stenopodidea. In: *Atlas of Crustacean Larvae*. Martin JW, Olesen J, Høeg JT (eds.). Baltimore, MD, USA: Johns Hopkins University Press, p. 243-249.

Groeneveld JC, Goñi R, Díaz D (2013) *Palinurus* species. In: *Lobsters: Biology, Management, Aquaculture and Fisheries* (2nd Edition). Phillips B. (ed.). West Sussex, UK: Wiley-Blackwell, p. 326-356.

Guerao G, Abelló P (1996) Descriptions of the first larval stage of *Polycheles typhlops* (Decapoda: Eryonidae: Polychelidae). *J Nat Hist* 30:1179-1184.

Gurney, R. (1942). *The Larvae of Decapod Crustacea*. London, UK: Ray Society, 306 p.

Hadley PB (1908) The behavior of the larval an adolescent stages of the American lobster (*Homarus americanus*). *J Comp Neurol Psychol* 18:199-301.

Harding GC, Drinkwater KF, Hannah CG, Pringle JD, Prena J, Loder JW, Pearre S Jr., Vass WP (2005) Larval lobster (*Homarus americanus*) distribution and drift in the vicinity of the Gulf of Maine offshore banks and their probable origins. *Fish Oceanogr* 14:112-137.

Harding GC, Pringle JD, Vass WP, Pearre S Jr., Smith SJ (1987) Vertical distribution and daily movements of larval lobsters *Homarus americanus* over Browns Bank, Nova Scotia. *Mar Ecol Prog Ser* 41:29-41.

Harvey A, Boyko CB, McLaughlin P, Martin JW (2014) Anomura. In: *Atlas of Crustacean Larvae*. Martin JW, Olesen J, Høeg JT (eds.). Baltimore, MD, USA: Johns Hopkins University Press, pp. 283-294.

Haye PA, Salinas P, Acuña E, Poulin E (2010) Heterochronic phenotypic plasticity with lack of genetic differentiation in the southeastern Pacific squat lobster *Pleuroncodes monodon*. *Evol Dev* 12:628-634.

Heasman KG, Jeffs AG (2019) Fecundity and potential juvenile production for aquaculture of the New Zealand Scampi, *Metanephrops challengeri* (Balss, 1914) (Decapoda: Nephropidae). *Aquaculture*. In Press. DOI:10.1016/j.aquaculture.2019.05.069.

Hill AE (1990) Pelagic dispersal of Norway lobster *Nephrops norvegicus* larvae examined using an advection-diffusion-mortality model. *Mar Ecol Prog Ser* 64:217-226.

Hill AE (1991) A mechanism for horizontal zooplankton transport by vertical migration in tidal currents. *Mar Biol* 111:485-492.

Holthuis LB (1991) *Marine Lobsters of the World. An Annotated and Illustrated Catalogue of the Species of Interest to Fisheries Known to Date*. Rome, Italy: Food and Agricultural Organization of the United Nations. FAO Species Catalogue No. 125, Vol. 13: p. 1-292.

Incze L, Xue H, Wolff N, Xu D, Wilson C, Steneck R, Wahle R, Lawton P, Pettigrew N, Chen Y (2010) Connectivity of lobster (*Homarus americanus*) populations in the coastal Gulf of Maine part II. Coupled biophysical dynamics. *Fish Oceanogr* 19:1-20.

Iwata Y, Sugita H, Deguchi Y, Kanemoto FI (1991) The larval development of reef lobster, *Enoplometopus occidentalis* Randall (Decapoda, Axiidae) reared in the laboratory. *Res Crustac* 20:1-15.

Iwata Y, Sugita H, Deguchi Y, Kamemoto FI (1992) On the larval morphology of *Metanephrops sagmariensis* reared in the laboratory. *Suisanzoshoku* 40:183-188.

Jackson DJ, MacMillan DL (2000) Tailflick escape behavior in larval and juvenile lobsters (*Homarus americanus*) and crayfish (*Cherax destructor*). *Biol Bull* 198:307-318.

Jeffs AG, Gardner C, Cockroft A (2013) *Jasus* and *Sagmariasus* species. In: *Lobsters: Biology, Management, Aquaculture and Fisheries* (2nd Edition). Phillips B. (ed.). West Sussex, UK: Wiley-Blackwell, p. 259-288.

Jirkowski GJ, Wolff C, Richter S (2015) Evolution of eumalacostracan development – new insights into loss and reacquisition of larval stages revealed by heterochrony analysis. *EvoDevo* 6:4. doi:10.1186/2041-9139-6-4.

Johnson M, Knight MW (1966) Phyllosoma larva of the spiny lobster *Panulirus inflatus* (Bouvier). *Crustaceana* 10:31-47.

Johnson ML, Johnson MP (eds.) (2013) The ecology and biology of *Nephrops norvegicus*. *Adv Mar Biol* 64:1-325.

Katz CH, Cobb JS, Spaulding M (1994) Larval behavior, hydrodynamic transport, and potential offshore-to-inshore recruitment in the American lobster *Homarus americanus*. *Mar Ecol Prog Ser* 103:265-272.

Kough AS, Paris CB, Butler MJ IV (2013) Larval connectivity and the international management of fisheries. *PLoS ONE* 8:e64970. doi:10.1371/journal.pone.0064970.

Kough AS, Paris CB, Staaterman E (2014) *In situ* swimming and orientation behavior of spiny lobster (*Panulirus argus*) postlarvae. *Mar Ecol Prog Ser* 504:207-219.

Kurata H (1965) Larvae of decapod Crustacea of Hokkaido, 9: Axiidae, Callianassidae, and Upogebiidae (Anomura). *Bull Hokkaido Reg Fish Res Lab* 30:1-10.

Lavalli KL, Spanier E (2010) Infraorder Palinura Latreille, 1802. In: *Treatise on Zoology, the Crustacea. Vol. 9A*. Schram F, von Vaupel Klein JC (eds.). Leiden, The Netherlands: Brill, p. 425-532.

Lawton P, Lavalli KL (1995) Postlarval, juvenile, adolescent, and adult ecology. In: *Biology of the Lobster* Homarus americanus. Factor JR (ed.). New York, NY, USA: Academic Press, Inc., p. 47-88.

Legall N, Poupin J (2019) Internet - CRUSTA: Database of Crustacea (Decapoda and Stomatopoda), with special interest for those collected in French overseas territories. Available at: http://crustiesfroverseas.free.fr/ (accessed 14 June 2019).

Lesser JHR (1978) Phyllosoma larvae of *Jasus edwardsii* (Hutton) (Crustacea: Decapoda: Palinuridae) and their distribution off the east coast of the North Island, New Zealand. *NZ J Mar Freshwat Res* 12:357-370.

Lohrmann KJ, Ernst DA (2013) The geomagnetic sense of crustaceans and its use in orientation and navigation. In: *The Natural History of Crustacea Volume 3: Nervous Systems and Control of Behavior*. Derby C, Thiel M (eds.). Oxford, UK: Oxford University Press, p. 321-336.

Lovrich GA, Thiel M (2011) Ecology, physiology, feeding and trophic role of squat lobsters. In: *Crustacean Issues 20: The Biology of Squat Lobsters*. Poore GCB, Ahyong ST, Taylor J (eds.). Collingwood, Australia: CRC Press, p. 183-222.

Macpherson W, Baba K (2011) Taxonomy of squat lobsters. In: *Crustacean Issues 20: The Biology of Squat Lobsters*. Poore GCB, Ahyong ST, Taylor J (eds.). Collingwood, Australia: CRC Press, p. 39-72.

Martin JW (2014a) Introduction to the Malacostraca. In: *Atlas of Crustacean Larvae*. Martin JW, Olesen J, Høeg JT (eds.). Baltimore, MD, USA: Johns Hopkins University Press, pp. 174-175.

Martin JW (2014b) Introduction to the Decapoda. In: *Atlas of Crustacean Larvae*. Martin JW, Olesen J, Høeg JT (eds.). Baltimore, MD, USA: Johns Hopkins University Press, pp. 230-234.

Martin JW (2014c) Polychelida. In: *Atlas of Crustacean Larvae*. Martin JW, Olesen J, Høeg JT (eds.). Baltimore, MD, USA: Johns Hopkins University Press, pp. 279-282.

Martin JW (2014d) Brachyura. In: *Atlas of Crustacean Larvae*. Martin JW, Olesen J, Høeg JT (eds.). Baltimore, MD, USA: Johns Hopkins University Press, p. 295-310.

Martin JW, Olesen J, Høeg JT (2014) Introduction. In: *Atlas of Crustacean Larvae*. Martin JW, Olesen J, Høeg JT (eds.). Baltimore, MD, USA: Johns Hopkins University Press, pp. 1-7.

Matsuda H, Yamakawa T (1997) Effects of temperature on growth of the Japanese spiny lobster, *Panulirus japonicus* (V. Siebold) phyllosomas under laboratory conditions. *Mar Freshwat Res* 48:791-796.

Medel C, Parada C, Morales CE, Pizarro O, Ernst B, Conejero C (2018) How biophysical interactions associated with sub- and mesoscale structures and migration behavior affect planktonic larvae of the spiny lobster in the Juan Fernández Ridge: A modeling approach. *Prog Oceanogr* 162:98-119.

Meyers ML, Jacobs MW, Gallager SM, Christmas AF (2012) Living it up before going down – vertical positioning behaviors of *Homarus americanus* larvae in response to a thermocline. In: *Society for Integrative and Comparative Biology 2012 Annual Meeting Abstracts. Integr Comp Biol* 52 (Suppl. 1):e295.

Newman BK, Papadopoulos I, Vorsatz J, Woolridge TH (2006) Influence of temperature on the larval development of *Upogebia africana* and *U. capensis* (Decapoda: Thalassinidae: Upogebiidae) in the laboratory. *Mar Ecol Prog Ser* 325:165-180.

Ngoc-Ho N (1977) The larval development of *Upogebia darwini* (Crustacea, Thalassinidea) reared in the laboratory, with a redescription of the adult. *J Zool* 181:439-464.

Okamoto K (2008) Japanese nephropid lobster *Metanephrops japonicus* lacks zoeal stage. *Fish Sci* 74:98-103.

Olesen J (2018) Crustacean life cycles – developmental strategies. In: *The Natural History of the Crustacea Volume 5: Life Histories*. Thiel M, Wellborn GA (eds.). Oxford, UK: Oxford University Press, p. 1-34.

Palero F, Clark PF, Guerao G (2014a) Achelata. In: *Atlas of Crustacean Larvae*. Martin JW, Olesen J, Høeg JT (eds.). Baltimore, MD, USA: Johns Hopkins University Press, pp. 272-278.

Palero F, Guerao G, Hall M, Chan TY, Clark PF (2014b) The 'giant phyllosoma' are larval stages of *Parribcus antarcticus* (Decapoda: Scyllaridae). *Invert System* 28:258-276.

Parker JJ (1884) On the structure of the head of *Palinurus* with special reference to the classification of the genus. *Trans New Zeland Inst* 16:297-307.

Perkins HC (1972) Developmental rates at various temperatures of embryos of the northern lobster (*Homarus americanus* Milne-Edwards). *Fish Bull* 70:95-99.

Phillips BF, Booth JD, Cobb JS, Jeffs AG, McWilliam P (2006) Larval and postlarval ecology. In: *Lobsters: Biology, Management, Aquaculture and Fisheries* (1st Edition). Phillips B. (ed.). Oxford, UK: Blackwell Publishing Ltd., p. 231-262.

Phillips BF, Cobb JS, George RW (1980) General biology. In: *The Biology and Management of Lobsters Volume I: Physiology and Behavior*. Cobb JS, Phillips BF (eds.). New York, NY, USA: Academic Press, Inc., p. 1-82.

Phillips BF, Cobb JS, Kittaka J (eds.) (1994) *Spiny Lobster Management*. Oxford, UK: Fishing News Books, Ltd., 550 p.

Phillips BF, Kittaka J (eds.) (2000) *Spiny Lobsters: Fisheries and Culture*. Oxford, UK: Blackwell Publishing Ltd., 704 p.

Phillips BF, Melville-Smith R, Kay MC, Vega-Velázquez A (2013a) *Panulirus* species. In: *Lobsters: Biology, Management, Aquaculture and Fisheries* (2nd Edition). Phillips B. (ed.). West Sussex, UK: Wiley-Blackwell, p. 289-325.

Phillips BF, Sastry AN (1980) Larval ecology. In: *The Biology and Management of Lobsters Volume II: Ecology and Management*. Cobb JS, Phillips BF (eds.). New York, NY, USA: Academic Press, Inc., p. 11-57.

Phillips BF, Wahle RA, Ward TJ (2013b) Lobsters as part of marine ecosystems – A review. In: *Lobsters: Biology, Management, Aquaculture and Fisheries* (2nd Edition). Phillips B. (ed.). West Sussex, UK: Wiley-Blackwell, p. 1-35.

Pike RB, Wear RG (1969) Newly hatched larvae of the genera *Gastroptychus* and *Uroptychus* (Crustacea, Decapoda, Galatheidea) from New Zealand waters. *Trans R Soc NZ* 11:189-195.

Pineda J, Reyns N (2018) Larval transport in the coastal zone: Biological and physical processes. In: *Evolutionary Ecology of Marine Invertebrate Larvae*. Carrier TJ, Reitzel AM, Heyland A (eds.). Oxford, UK: Oxford University Press, p. 141-159.

Pohle G, Santana W (2014) Gebiidea and Axiidea (= Thalassinidea). In: *Atlas of Crustacean Larvae*. Martin JW, Olesen J, Høeg JT (eds.). Baltimore, MD, USA: Johns Hopkins University Press, pp. 263-271.

Pohle G, Santana W, Jansen G, Greenlaw M (2011) Plankton-caught zoeal stages and megalopa of the lobster shrimp *Axius serratus* (Decapoda: Axiidae) from the Bay of Fundy, Canada, with a summary of axiidean and gebiidean literature on larval descriptions. *J Crustac Biol* 31:82-99.

Poulsen EM (1946) Investigations on the Danish fishery for and the biology of the Norway lobster and the deep-sea prawn. *Rep Dan Biol Stn* 48:27-49.

Quinn BK (ed.) (2019) *Lobsters: Biology, Behavior and Management*. New York, NY, USA: Nova Science Publishers, Inc., 172 p.

Quinn BK, Chassé J, Rochette R (2017) Potential connectivity among American lobster fisheries as a result of larval drift across the species' range in eastern North America. *Can J Fish Aquat Sci* 74:1549-1563.

Roa R, Gallardo VA, Ernst B, Baltazar M, Cañete JI, Enriquez-Briones S (1995) Nursery ground, age structure and abundance of juvenile squat lobster *Pleuroncodes monodon* on the continental shelf off central Chile. *Mar Ecol Progr Ser* 116:47-54.

Robertson PB (1968) *The Larval Development of Some Western Atlantic Lobsters of the Family Scyllaridae*. PhD Thesis. University of Miami, Coral Gables, FL, USA.

Robertson PB (1969) Biological investigations of the deep sea, No. 48. Phyllosoma larvae of the scyllarid lobster, *Arctides guineensis*, from the western Atlantic. *Mar Biol* 4:143-151.

Robertson PB (1971) Biological results of the University of Miami deep-sea expeditions. 84. The larvae and postlarva of the scyllarid lobster *Scyllarus depressus* (Smith). *Bull Mar Sci* 21:841-865.

Saigusa M, Hirano Y, Kang BJ, Sekiné K, Hatakeyama M, Nanri T, Hamaguchi M, Masunari N (2018) Classification of the intertidal and estuarine upogebiid shrimps (Crustacea: Thalassinidea), and their settlement in the Ryukyu Islands, Japan. *J Mar Biol Oceanogr* 7:2. DOI:10.4172/2324-8661.1000192.

Sakai K (2004) The diphyletic nature of the infraorder Thalassinidea (Decapoda, Pleocyemata) as derived from the morphology of the gastric mill. *Crustaceana* 77:1117-1129.

Schmalenbach I, Buchholtz F (2010) Vertical positioning and swimming performance of lobster larvae (*Homarus gammarus*) in an artificial water column at Helgoland, North Sea. *Mar Biol Res* 6:89-99.

Schmalenbach I, Franke HD (2010) Potential impact of climate warming on the recruitment of an economically and ecologically important species, the European lobster (*Homarus gammarus*) at Helgoland, North Sea. *Mar Biol* 157:1127-1135.

Segura-Garciá I, Garavelli L, Tringali M, Matthews T, Chérubin LM, Hunt J, Box SJ (2019) Reconstruction of larval origins based on genetic relatedness and biophysical modeling. *Sci Rep* 9:7100. DOI:10.1038/s41598-019-43435-9.

Sekiguchi H (1988) Taxonimcal and ecological problems associated with phyllosoma larvae. *Benthos Res* (*Bull Jap Assoc Benthology*) 33/34:1-16. [In Japanese, abstract in English.]

Sekiguchi H, Booth JD, Webber WR (2007) Early life histories of slipper lobsters. In: *Crustacean Issues 17: The Biology and Fisheries of the Slipper Lobster*. Lavalli KL, Spanier E (eds.). Boca Raton, FL, USA: CRC Press, p. 69-90.

Sekiguchi H, Inoue N (2002) Recent advances in larval recruitment processes of scyllarid and palinurid lobsters in Japanese waters. *J Oceanogr* 58:747-757.

Serfing SA, Ford RF (1975) Ecological studies of the puerulus larval stage of the California spiny lobster, *Panulirus interruptus*. *Fish Bull* 73:360-367.

Sims HW Jr. (1964) Four giant scyllarid phyllosoma larvae from the Florida Straits with notes on smaller similar specimens. *Crustaceana* 7:259-266.

Sims HW Jr. (1965) Notes on the occurrence of prenaupliosoma larvae of spiny lobsters in the plankton. *Bull Mar Sci* 15:223-227.

Sims HW Jr. (1966) The phyllosoma larvae of the spiny lobster *Palinurellus gundlachi* Von Martens (Decapoda, Palinuridae). *Crustaceana* 11:205-215.

Singh SP, Groeneveld JC, Willows-Munro S (2019) Between the current and the coast: Genetic connectivity in the spiny lobster *Panulirus homarus rubellus*, despite potential barriers to gene flow. *Mar Biol* 166:36. DOI:10.1007/s00227-019-3486-4.

Spanier E, Lavalli K (2007) Slipper lobster fisheries – present status and future perspectives. In: *Crustacean Issues 17: The Biology and Fisheries of the Slipper Lobster*. Lavalli KL, Spanier E (eds.). Boca Raton, FL, USA: CRC Press, p. 377-392.

Spanier E, Lavalli KL (2013) Commercial scyllarids. In: *Lobsters: Biology, Management, Aquaculture and Fisheries* (2nd Edition). Phillips B. (ed.). West Sussex, UK: Wiley-Blackwell, p. 414-466.

Stanley RRE (2015) *Laboratory- and Field-Based Approaches for Evaluating Connectivity in a Dynamic Coastal Environment: Applications for Management and Conservation*. PhD Thesis, Memorial University of Newfoundland, St. John's, NL, Canada.

Stanley RRE, Pedersen EJ, Snelgrove PVR (2016) Biogeographic, ontogenetic, and environmental variability in larval behaviour of American lobster *Homarus americanus*. *Mar Ecol Prog Ser* 553:125-146.

Strathmann RR (2018) Larvae and direct development. In: *The Natural History of the Crustacea Volume 5: Life Histories*. Thiel M, Wellborn GA (eds.). Oxford, UK: Oxford University Press, p. 151-178.

Tamaki A, Tanoue H, Itoh J, Fukuda Y (1996) Brooding and larval developmental periods of callianassid ghost shrimp, *Callianassa japonica* (Decapoda: Thalassinidea). *J Mar Biol Assoc UK* 76:675-689.

Templeman W (1936) The influence of temperature, salinity, light and food conditions on the survival and growth of the larvae of the lobster (*Homarus americanus*). *J Biol Board Can* 2:485-497.

Ting RY (1973) Culture potential of spiny lobster. *Proc Annu Meet World Maric Soc* 4:165-270.

Torres AP, Guerao G, Abelló P, Blanco E, Boné A, Palero F, Dos Santos A, Hidalgo M (2013) Vertical distribution in relation with food availability and morphology of the larval stages of *Polycheles typhlops* (Decapoda, Polychelida). *ICES Conference and Meeting Document* 2013/F:14. Available online at: http://www.ices.dk/sites/pub/CM%20Doccuments/CM-2013/Theme%20Session%20F%20contributions/F1413.pdf (accessed 12 June 2019).

Torres AP, Palero F, Dos Santos A, Abelló P, Blanco E, Boné A, Guerao G (2014) Larval stages of the deep-sea lobster *Polycheles typhlops* (Decapoda, Polychelida) identified by DNA analysis: Morphology, systematic, distribution and ecology. *Helgol Mar Res* 68:379-397.

Tudge CC, Asakura A, Ahyong ST (2012) Infraorder Anomura MacLeay, 1838. In: Schram F, von Vaupel Klein JC (eds.). *Treatise on Zoology, the Crustacea. Vol. 9B*. Leiden, The Netherlands: Brill, p. 221-333.

Tully O, Ó Céidigh P (1987) The seasonal and diel distribution of lobster larvae (*Homarus gammarus* (Linnaeus)) in the neuston of Galway Bay. *J Cons int Explor Mer* 44:5-9.

Uchida T, Dotsu Y (1973) Collection of the T. S. *Nagasaki Maru* of Nagasaki University, 4: on the larval hatching and larval development of the lobster *Nephrops thomasoni*. *Bull Fac Fish Nagasaki Univ* 36:23-35.

Vogt G, Tolley L (2004) Brood care in freshwater crayfish and relationship with the offspring's sensory deficiencies. *J Morphol* 262:566-582.

Wahle RA, Castro KM, Tully O, Cobb JS (2013) *Homarus*. In: *Lobsters: Biology, Management, Aquaculture and Fisheries* (2nd Edition). Phillips B. (ed.). West Sussex, UK: Wiley-Blackwell, p. 221-258.

Wahle RA, Tshudy D, Cobb JS, Factor J, Jaini M (2012) Infraorder Astacidea Latreille, 1802 p.p.: The marine clawed lobsters. In: *Treatise on Zoology, the Crustacea. Vol. 9B*. Schram F, von Vaupel Klein JC (eds.). Leiden, The Netherlands: Brill, p. 3-107.

Wakabayashi K, Tanaka Y (2012) The jellyfish-rider: phyllosoma larvae of spiny and slipper lobsters associated with jellyfish. *TAXA* 33:5-12. [In Japanese, abstract in English.]

Wear RG (1976) Studies on the larval development of *Metanephrops challengeri* (Balss, 1914) (Decapoda, Nephropidae). *Crustaceana* 30:113-122.

Wehrtmann IS, Acuña E (2011) Squat lobster fisheries. In: *Crustacean Issues 20: The Biology of Squat Lobsters*. Poore GCB, Ahyong ST, Taylor J (eds.). Collingwood, Australia: CRC Press, p. 297-322.

Wilkens H, Parzefall J, Ribowski A (1990) Population biology and larvae of the anchialine crab *Munidopsis polymorpha* (Galatheidae) from Lanzarote (Canary Islands). *J Crustac Biol* 10:667-675.

Williams AB (1986) Lobsters – Identification, world distribution, and U.S. trade. *Mar Fish Rev* 48:1-36.

Williams AB (1988) *Lobsters of the World – An Illustrated Guide*. New York, NY, USA: Osprey Books, 186 p.

Williamson DI (1982) Larval morphology and diversity. In: *The Biology of the Crustacea, Vol. 2: Embryology, Morphology, and Genetics*. Abele LG (ed.). New York, NY, USA: Academic Press, p. 43-110.

Williamson DI (1983) Crustacea Decapoda: Larvae VIII. Nephropidea, Palinuridea, and Eryonidea. *Fich Ident Zooplancton* 167/168:8 p.

Williamson DI, Rice AL (1966) Larval evolution in the Crustacea. *Crustaceana* 69:267-287.

Wolfe JM, Breinholt JW, Crandall KA, Lemmon AR, Moriarty Lemmon E, Timm LE, Siddall ME, Bracken-Grissom D (2019) A phylogenomic framework, evolutionary timeline and genomic resources for comparative studies of decapod crustaceans. *Proc R Soc B* 286. DOI:10.1098/rspb.2019.0079.

Xue H, Incze L, Xu D, Wolff N, Pettigrew N (2008) Connectivity of lobster populations in the coastal Gulf of Maine part I: Circulation and larval transport potential. *Ecol Model* 210:193-211.

Yanagimoto T, Konishi K, Takami M, Saruwatari T (2015) Species identification of polychelid postlarvae collected from the Pacific coast of Honshu, Japan using DNA analyses. *Cancer* 24:7-13. [In Japanese, abstract in English.].

INDEX

A

abbreviated development, 123, 125, 138
abdominal spines, 125, 128, 130, 132, 135
Achelata, ix, 113, 114, 115, 117, 118, 123, 127, 129, 130, 150, 157
adults, 116, 117, 118, 133, 134, 139, 142, 143, 144, 147
advanced development, 128, 132, 135
air temperature, 71, 72
Al Wathba Wetland Reserve, viii, 68, 70, 71, 72, 74, 75, 76, 84, 87
alarm cues, ix, 90, 91, 93, 94, 96, 97, 99, 105, 106, 111
Anomura, ix, 113, 117, 118, 124, 129, 130, 151, 152, 154, 155, 162
antipredator behaviour, 90, 106
apex, 11, 16, 22, 23, 25, 26, 42, 44, 51, 52, 53, 54, 55
aquaculture, viii, 67, 86, 154
aquarium, 93, 94
aquatic habitats, 106
aquatic systems, 92
Artemia, v, vii, viii, 67, 68, 69, 74, 75, 76, 83, 84, 85, 86, 87, 88

Artemia franciscana, v, vii, viii, 67, 68, 74, 75, 76, 85, 87, 88
articulation, 28, 30, 40, 42, 44, 47, 48, 49
Astacidea, ix, 113, 114, 115, 117, 118, 122, 123, 127, 129, 130, 149, 150, 152, 153, 163
Astacoidea, 115, 117, 122, 130, 131
avian, viii, 68, 69, 84, 86, 87
Axiidea, ix, 113, 114, 116, 117, 124, 129, 130, 131, 132, 138, 151, 159

B

behavioral change, 121
behaviors, vii, x, 114, 120, 126, 138, 140, 143, 144, 147, 148, 157
benthic recruitment, 119
biodiversity, 60
birds, viii, 68, 70, 80, 83, 84, 85, 87, 92
brine shrimps, 68, 69, 70, 71, 80, 82, 83, 84, 85

C

Callianassidae, 124, 155

capsule, 77, 78, 79, 82
carapace, 11, 12, 13, 15, 18, 19, 22, 23, 24, 26, 28, 30, 36, 37, 41, 42, 44, 48, 49, 50, 51, 52, 53, 54, 56, 58, 125, 134, 135
Caribbean, 6, 61, 65, 142, 153
cellulose, 94, 95, 97, 98, 99, 100, 101, 102
cephalothorax, 114, 125, 132, 134
cestodes, v, viii, 67, 68, 69, 81, 82, 83, 85, 86, 87
chelae, 10, 11, 17, 31, 53, 57, 115, 128, 131
cheliped, 16, 28, 41
chemical, vii, ix, 90, 91, 92, 93, 94, 95, 103, 104, 106, 109
chemical cues, v, ix, 89, 90, 91, 93, 94, 96, 97, 99, 103, 104, 108, 109
Chirostyloidea, ix, 113, 114, 117, 118, 124, 129, 135, 138
clawed lobster, 115, 122, 123, 136, 137, 140, 142, 145, 163
climate change, 120
coastal communities, 118
coastal region, 140
compounds, vii, ix, 90, 91, 93, 95, 103, 106
connectivity, vii, x, 65, 114, 120, 141, 148, 151, 154, 155, 159, 161, 164
control group, 95, 97, 99, 100, 102, 104, 105
coral reefs, 115, 116
cornea, 11, 22, 28, 30, 33, 35, 37, 40, 42, 44, 46, 48, 49, 52, 53, 55
crabs, ix, 2, 60, 63, 111, 113, 118, 124, 125, 135, 144
crayfish, viii, 50, 62, 67, 117, 122, 131, 151, 152, 155, 162
cues, ix, 90, 91, 93, 94, 96, 103, 104, 105, 106, 107, 108, 109, 111, 140, 142, 145, 146
cysticercoids, viii, 68, 70, 71, 73, 74, 75, 76, 81, 82, 83, 84, 85, 88

D

decapodid, x, 114, 121, 122, 123, 124, 128, 130, 131, 132, 133, 135, 137, 138, 139, 141, 144, 146, 147
defence, 90, 104, 111
defensive strategies, ix, 90, 91
depression, 7, 30, 40, 42, 44, 47, 52, 58
Dhabi, Abu, v, viii, 67, 68, 70, 71, 72, 74, 75, 84, 87, 88
distribution, viii, 1, 35, 61, 91, 92, 97, 100, 102, 103, 110, 144, 149, 153, 154, 156, 162, 163
diversity, vii, x, 2, 58, 103, 114, 132, 135, 163

E

ecological roles, ix, 113
ecology, ix, 86, 106, 108, 109, 110, 114, 120, 123, 147, 152, 155, 158, 162
ecosystem, 109
egg, 121, 125, 131, 152
Enoplometopoidea, ix, 113, 115, 117, 122, 127, 129, 130
environmental stimuli, 142
environmental stress, 92, 112
environments, vii, viii, ix, 1, 2, 59, 60, 66, 90, 91, 105, 142
eryoneicus, x, 114, 116, 124, 129, 130, 134, 137, 143
evidence, 112, 139, 142, 143, 145, 146, 152
exopodite, 27, 36, 139
experimental condition, 111

F

families, vii, viii, 2, 8, 58, 59, 68
fauna, 3, 8, 61, 63, 85, 86, 87
fish, viii, 67, 69, 85, 92

fisheries, vii, ix, 110, 113, 118, 119, 122, 148, 149, 153, 154, 155, 158, 159, 160, 161, 162, 163
flagellum, 12, 18, 22, 23, 25, 26, 28, 31, 36, 38, 41, 42, 44, 49, 57
flexor, 10, 11, 14, 17, 19, 21, 32
food, viii, 2, 59, 67, 70, 80, 81, 84, 92, 93, 103, 108, 109, 111, 124, 143, 162
freshwater, vii, viii, 1, 2, 15, 17, 21, 27, 35, 37, 39, 41, 44, 46, 47, 48, 49, 50, 51, 52, 54, 57, 59, 60, 61, 63, 64, 105, 108, 109, 110, 115, 117, 122, 131, 152, 162
furry lobster, ix, 113, 116, 123, 148

G

Galatheoidea, ix, 113, 114, 117, 118, 124, 129, 130, 135, 138
Gebiidea, ix, 113, 114, 116, 117, 124, 127, 129, 130, 131, 132, 138, 151, 159
genus, viii, 57, 60, 63, 64, 65, 67, 68, 70, 73, 81, 82, 88, 109, 133, 151, 158
geology, 2
glaucothoe, 124, 130, 135
glycerin, vii, viii, 68, 71
Glypheidea, ix, 113, 116, 117, 123, 150
granules, 36, 52, 56
growth, 152, 153, 157, 162
Guatemala, 3, 6, 7, 60, 63, 65, 66
Gulf Coast, 5
Gulf of Mexico, 4

H

habitats, vii, viii, 1, 92, 105, 115, 119, 120, 125, 139, 140, 141, 142, 143, 144, 146, 147
Heterocypris incongruens, vii, ix, 90, 92, 94, 108, 109, 110, 111, 112

Homarus, 114, 117, 118, 122, 125, 127, 128, 129, 130, 136, 145, 149, 150, 151, 152, 153, 154, 155, 157, 158, 160, 161, 162
horizontal swimming, 128, 139, 144, 145, 146, 147, 148

I

individuals, ix, 16, 86, 90, 91, 93, 94, 96, 97, 98, 99, 100, 101, 102, 104, 105, 106
infection, 73, 84, 86, 87
infochemicals, 91, 103, 104, 106
insects, 92, 109
insertion, 52, 53, 56
instar, 121, 123, 130
ischium, 13, 14, 16, 18, 21, 22, 23, 25, 27, 28, 30, 32, 34, 36, 38, 40, 45, 47, 52, 53, 54, 57, 58

J

Jasus, 118, 126, 127, 129, 137, 142, 151, 155, 156

K

Kiwa, 117, 118, 125

L

lakes, 7, 80, 82, 88
lamella, 26, 38
Laomediidae, 129, 131, 152
larva, 80, 83, 88, 114, 116, 125, 134, 140, 155
larvae, viii, ix, x, 67, 71, 73, 83, 88, 90, 92, 93, 94, 107, 111, 114, 115, 116, 119, 120, 121, 123, 124, 125, 128, 131, 132, 134, 135, 136, 137, 138, 139, 140, 142,

143, 144, 145, 147, 148, 149, 150, 152, 153, 154, 156, 157, 158, 159, 160, 161, 162, 163
larval development, vii, x, 114, 120, 123, 124, 129, 134, 136, 148, 149, 151, 152, 154, 157, 159, 161, 162, 163
larval dispersal, vii, x, 114, 120, 141, 143, 144, 146, 148
larval stages, viii, 67, 82, 120, 122, 123, 124, 128, 131, 138, 140, 141, 143, 144, 150, 155, 157, 162
life cycle, viii, ix, 2, 68, 70, 114, 119, 121, 123, 146, 157
life expectancy, 69, 81
limestone, 3, 4, 5, 6, 7
lobster, vii, ix, 114, 116, 117, 118, 119, 120, 121, 122, 125, 126, 127, 128, 129, 130, 131, 132, 135, 136, 137, 139, 140, 142, 144, 145, 146, 147, 148, 149, 150, 151, 152, 153, 154, 155, 157, 158, 159, 160, 161, 162, 164

M

marine environment, 120
maxilliped, 11, 12, 15, 17, 19, 21, 22, 23, 25, 35, 47, 52, 53, 54, 55, 57, 58
median, 12, 26, 28, 34, 35, 37, 48, 49, 52, 53, 54, 56
megalopa, 124, 130, 132, 159
metamorphosis, 121, 123, 133
Metanephrops, 122, 136, 154, 157, 163
Mexico, v, vii, 1, 2, 4, 6, 7, 8, 15, 41, 44, 46, 48, 54, 59, 60, 61, 62, 63, 64, 65
microscopy, vii, viii, 68
migration, 104, 142, 154, 157
morphology, ix, 86, 103, 109, 112, 114, 116, 120, 124, 125, 133, 154, 160, 162, 163
mortality, 90, 139, 147, 154
mountain ranges, 3
mud lobster, ix, 113, 116, 124
mysis, x, 114, 121, 126

N

natatory setae, 125, 132
National Academy of Sciences, 110
National Oceanic and Atmospheric Administration, 117
naupliosoma, 126
Nephropoidea, ix, 113, 115, 117, 118, 122, 127, 129, 130
Nephrops, 118, 122, 128, 136, 149, 154, 155, 162
Netherlands, 149, 151, 153, 155, 162, 163
New Zealand, 154, 156, 158
nisto, 123, 130, 133, 137, 146
North America, 107, 118, 159
Norway, 122, 154, 159
Nuevo León, 61, 63

O

orbit, 28, 35, 37, 54, 57
organs, 55, 146, 152
Ostracoda, v, ix, 89, 90, 107, 108, 109, 110, 111, 112
oxygen, 59, 69, 119
oxygen consumption, 59
oxygen consumption rate, 59

P

Pacific, 4, 5, 6, 151, 154, 164
Palinurellus, 117, 123, 129, 151, 161
Palinuridae, ix, 113, 115, 117, 123, 127, 129, 130, 151, 156, 161
Panulirus, 65, 117, 118, 130, 137, 142, 146, 150, 153, 155, 157, 158, 160, 161
parasites, 82, 86, 87

Parastacoidea, 115, 122, 131
pelagic larval duration (PLD), 136, 137, 138, 148
pereiopod, 10, 11, 13, 14, 16, 19, 20, 22, 23, 25, 27, 30, 32, 34, 35, 36, 38, 40, 47, 48, 49, 50, 51, 58
Phyllamphion, 129, 133
phyllosoma, x, 114, 116, 123, 126, 129, 132, 137, 142, 149, 151, 155, 156, 157, 159, 160, 161, 163
physical structure, 119
physiology, ix, 103, 114, 156
plankton, 109, 126, 133, 136, 137, 139, 142, 143, 161
pleopod, 11, 13, 15, 17, 21, 22, 24, 25, 27, 33, 34, 35, 36, 39, 48, 50, 51, 57, 132, 135, 139
pleura, 18, 22, 23, 24, 26, 28, 29, 31, 33, 35, 37, 39, 41, 42, 44, 46
Pleuroncodes, 118, 138, 144, 147, 153, 154, 159
Polycheles, 127, 143, 153, 162
Polychelida, ix, 113, 114, 116, 117, 124, 127, 129, 130, 150, 156, 162
ponds, 7, 15, 92, 110, 112
population, viii, 68, 84, 93, 111, 119, 147, 153
population structure, 153
postlarva, x, 114, 121, 122, 159
predation, 90, 91, 105, 108, 111, 127
predator, vii, ix, 90, 91, 92, 93, 94, 95, 97, 98, 99, 100, 101, 102, 103, 104, 105, 106, 108, 109, 110, 111, 138
pre-zoea, x, 114, 121, 125, 126, 127, 148
puerulus, 123, 130, 133, 137, 146, 150, 160

Q

quartz, 94, 95, 97, 98, 99, 100, 101, 102, 103, 105

R

reef lobsters, ix, 113, 115, 122, 131, 148
response, 91, 95, 103, 106, 140, 157
rock lobster, 115, 118, 150, 151
room temperature, 93

S

salinity, 69, 70, 80, 84, 85, 92, 140, 145, 162
Scyllaridae, ix, 113, 115, 117, 123, 129, 130, 152, 157, 159
sea level, 5, 7
seafood, 118
sediment, 144
settlement, 119, 120, 121, 122, 124, 128, 134, 136, 137, 144, 145, 146, 147, 149, 150, 151, 152, 160
shrimp, ix, 2, 13, 23, 27, 29, 33, 35, 37, 41, 44, 46, 47, 61, 63, 65, 68, 70, 73, 84, 85, 87, 88, 114, 131, 135, 152, 159, 161
Silentes, 133
slipper lobster, 115, 118, 123, 160, 161, 163
South Africa, 65, 149
Spain, 82, 83, 85, 87
species, vii, viii, 1, 8, 9, 12, 17, 19, 21, 23, 24, 25, 27, 29, 30, 33, 35, 39, 46, 48, 50, 51, 52, 55, 58, 59, 60, 61, 62, 63, 64, 65, 68, 69, 70, 71, 73, 75, 76, 80, 81, 82, 83, 84, 85, 86, 87, 91, 92, 105, 107, 109, 115, 116, 118, 119, 122, 123, 124, 126, 130, 132, 134, 135, 136, 137, 138, 142, 144, 146, 147, 149, 151, 153, 155, 158, 159, 160
spine, 14, 16, 18, 20, 23, 24, 26, 28, 29, 31, 34, 35, 37, 39, 42, 45, 46, 47, 48, 49, 50, 51, 52, 53, 56, 58
spiny lobster, 65, 142, 150, 151, 153, 155, 157, 158, 160, 161, 162
sponge, 94, 95, 96, 97, 98, 99, 100, 101, 102

squat lobster, ix, 113, 118, 124, 125, 131, 135, 138, 144, 147, 148, 149, 154, 156, 159, 163
sternum, 14, 16, 20, 48
Stridentes, 126, 127, 133
structure, 82, 106, 116, 120, 133, 158, 159
subacute, 12, 13, 20, 25, 42, 44
substrate, ix, 90, 92, 93, 105, 136
survival, ix, 59, 85, 90, 92, 104, 105, 119, 138, 152, 162
suture, 13, 15, 20, 28, 52, 58
Synaxidae, ix, 113, 115, 117, 129

T

taxa, x, 91, 114, 117, 121, 123, 124, 125, 127, 129, 130, 132, 133, 135, 144, 147, 148
teeth, 11, 27, 28, 29, 33, 34, 35, 37, 39, 41, 43, 44, 46, 47, 52, 54, 55, 56, 57
telson, 12, 14, 16, 18, 20, 22, 23, 24, 26, 28, 29, 31, 33, 34, 35, 37, 40, 41, 42, 44, 46, 47, 49, 131, 135
temperature, 69, 71, 72, 80, 84, 85, 124, 136, 140, 149, 150, 157, 162
Thalassinidea, ix, 113, 116, 117, 124, 127, 129, 130, 131, 151, 152, 157, 159, 160, 161
thalassinidean hair, 131, 132
tooth, 11, 12, 22, 23, 24, 26, 28, 29, 31, 33, 34, 37, 50, 51, 52, 53, 55, 57
transport, 150, 154, 155, 159, 164

U

United Arab Emirates, v, viii, 67, 68, 70, 84, 87, 88
Upogebia, 117, 130, 138, 157
uropod, 12, 15, 17, 21, 26, 34, 36, 42, 44, 135
USA, 82, 152, 153, 154, 155, 156, 157, 158, 159, 160, 161, 163

V

vertical migratory behavior, 114
vertical positioning, 139, 140, 142, 144, 157, 160

W

water, viii, x, 68, 70, 71, 72, 73, 80, 91, 92, 93, 94, 95, 96, 104, 108, 109, 114, 115, 118, 120, 121, 123, 136, 139, 140, 142, 143, 146, 147, 151, 160
Western Australia, 150

Y

yeti crab, 118, 125

Z

zoea, x, 114, 121, 122, 124, 125, 126, 128, 129, 131
zooplankton, 133, 143, 154

Related Nova Publications

LIVESTOCK: PRODUCTION, MANAGEMENT STRATEGIES AND CHALLENGES

EDITORS: Victor Roy Squires and Wayne L. Bryden

SERIES: Animal Science, Issues and Research

BOOK DESCRIPTION: This book brings together and discusses information relating to animal production systems in different parts of the world. Throughout this book there are examples of systems comprised of a collection of interdependent and interactive elements that act together to accomplish a desired outcome.

HARDCOVER ISBN: 978-1-53615-540-2
RETAIL PRICE: $270

AMPHIBIANS: BIOLOGY, ECOLOGY AND CONSERVATION

EDITOR: Leo Cannon

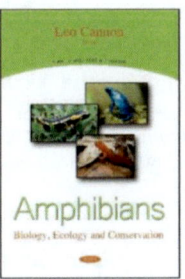

SERIES: Animal Science, Issues and Research

BOOK DESCRIPTION: *Amphibians: Biology, Ecology and Conservation* opens with a review of the current knowledge on the Harderian gland of anuran amphibians.

SOFTCOVER ISBN: 978-1-53614-034-7
RETAIL PRICE: $82

To see a complete list of Nova publications, please visit our website at www.novapublishers.com

Related Nova Publications

ZOONOMIA. VOLUME I: THE LAWS OF ORGANIC LIFE

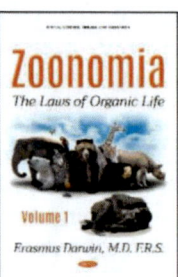

AUTHOR: Erasmus Darwin, M.D.

SERIES: Animal Science, Issues and Research

BOOK DESCRIPTION: This 2 volume set is written by the grandfather of Charles Darwin. The purpose of the books is to reduce the facts belonging to animal life into classes, orders, genera, and species; and, by comparing them with each other, to unravel the theory of diseases.

HARDCOVER ISBN: 978-1-53614-872-5
RETAIL PRICE: $250

ZOONOMIA. VOLUME II: THE LAWS OF ORGANIC LIFE

AUTHOR: Erasmus Darwin, M.D.

SERIES: Animal Science, Issues and Research

BOOK DESCRIPTION: This 2 volume set is written by the grandfather of Charles Darwin. The purpose of the books is to reduce the facts belonging to animal life into classes, orders, genera, and species; and, by comparing them with each other, to unravel the theory of diseases.

HARDCOVER ISBN: 978-1-53614-874-9
RETAIL PRICE: $275

To see a complete list of Nova publications, please visit our website at www.novapublishers.com